Gas-Engines and Producer-Gas Plants

A PRACTICE TREATISE SETTING FORTH THE PRINCIPLES OF GAS-ENGINES AND PRODUCER DESIGN, THE SELECTION AND INSTALLATION OF AN ENGINE, CONDITIONS OF PERFECT OPERATION, PRODUCER-GAS ENGINES AND THEIR POSSIBILITIES, THE CARE OF GAS-ENGINES AND PRODUCER-GAS PLANTS, WITH A CHAPTER ON VOLATILE HYDROCARBON AND OIL ENGINES

BY

R. E. MATHOT, M.E.

Member of the Société des Ingénieurs Civils de France, Institution of Mechanical Engineers, Association des Ingénieurs de l'Ecole des Mines du Hainaut of Brussels

TRANSLATED FROM ORIGINAL FRENCH MANUSCRIPT BY

WALDEMAR B. KAEMPFFERT

WITH A PREFACE BY

DUGALD CLERK, M. INST. C.E., F.C.S.

ILLUSTRATED

NEW YORK
MUNN & COMPANY
OFFICE OF THE SCIENTIFIC AMERICAN
361 BROADWAY
1905

Gas-Engines and Producer-Gas Plants

by R. E. Mathot

Originally published by
Munn & Company
New York

Reprinted by
Lindsay Publications Inc
Bradley IL 60915

ISBN 1-55918-245-8

2 3 4 5 6 7 8 9 0

2001

WARNING

Remember that the materials and methods described here are from another era. Workers were less safety conscious then, and some methods may be downright dangerous. Be careful! Use good solid judgement in your work, and think ahead. Lindsay Publications Inc. has not tested these methods and materials and does not endorse them. Our job is merely to pass along to you information from another era. Safety is your responsibility.

PREFACE

TO

"MATHOT'S GAS-ENGINES AND PRODUCER-GAS PLANTS"

BY

DUGALD CLERK, M.INST.C.E., F.C.S.

MR. MATHOT, the author of this interesting work, is a well-known Belgian engineer, who has devoted himself to testing and reporting upon gas and oil engines, gas producers and gas plants generally for many years past. I have had the pleasure of knowing Mr. Mathot for many years, and have inspected gas-engines with him. I have been much struck with the ability and care which he has devoted to this subject. I know of no engineer more competent to deal with the many minute points which occur in the installation and running of gas and oil engines. I have read this book with much interest and pleasure, and I consider that it deals effectively and fully with all the principal detail points in the installation, operation, and testing of these engines. I know of no work which has gone so fully into the details of gas-engine installation and up-keep. The work clearly points out all the matters which have to be attended to in getting the best work from any gas-engine under the varying circumstances of different installations and conditions. In my view, the book is a most useful one, which deserves, and no doubt will obtain, a wide public recognition.

DUGALD CLERK.

March, 1905.

INTRODUCTION

THE constantly increasing use of gas-engines in the last decade has led to the invention of a great number of types, the operation and care of which necessitate a special practical knowledge that is not exacted by other motors, such as steam-engines.

Explosion-engines, driven by illuminating-gas, producer-gas, oil, benzin, alcohol and the like, exact much more care in their operation and adjustment than steam-engines. Indeed, steam-engines are regularly subjected to comparatively low pressures. The temperature in the cylinders, moreover, is moderate.

On the other hand, the explosion-motor is irregularly subjected to high and low pressures. The temperature of the gases at the moment of explosion is exceedingly high. It is consequently necessary to resort to artificial means for cooling the cylinder; and the manner in which this cooling is effected has a very great influence on the operation of the motor. If the cooling be effected too rapidly, the quantity of gas consumed is considerably increased; if the cooling be effected too slowly, the motor parts will quickly deteriorate.

In order to reduce the gas consumption to a minimum, a matter which is particularly important when the motor is driven by street-gas, the explosive mixture is compressed before ignition. Only if all the parts are built with joints absolutely gas-tight is it possible to obtain this compression. The slightest leakage past the valves or around the piston will sensibly increase the consumption.

The mixture should be exploded at the exact moment the piston starts on its working stroke. If ignition occur too soon or too late, the result will be a marked diminution in the useful effect produced by the expansion of the gas. All ignition devices are composed of delicate parts, which cannot be too well cared for.

It follows from what has thus far been said that the causes of perturbation are more numerous in a gas than in a steam engine; that with a gas-engine, improper care will lead to a much greater increase in consumption than with a steam-engine, and will cause a waste in power which would hardly be appreciable in steam-engines, whether their joints be tight or not.

It is the purpose of this manual to indicate the more elementary precautions to be taken in the care of an engine operating under normal conditions, and to explain how repairs should be made to remedy the injuries caused by accidents. Engines which are of less than 200 horse-power and which are widely used in a small way will be primarily considered. In another work the author will discuss more powerful engines.

Before considering the choice, installation, and operation of a gas-engine, it will be of interest to ascertain the relative cost of different kinds of motive power. Disregarding special reasons which may favor the one or the other method of generating power, the net cost per horse-power hour will be considered in each case in order to show which is the least expensive method of generating power in ordinary circumstances.

R. E. MATHOT.

MARCH, 1905.

TABLE OF CONTENTS

CHAPTER V

WATER CIRCULATION

CHAPTER VI

LUBRICATION

CHAPTER VII

CONDITIONS OF PERFECT OPERATION

CHAPTER VIII

HOW TO START AN ENGINE.—PRELIMINARY PRECAUTIONS

CHAPTER IX

PERTURBATIONS IN THE OPERATION OF ENGINES AND THEIR REMEDY

CHAPTER X

PRODUCER-GAS ENGINES

CHAPTER XI

PRODUCER-GAS

CHAPTER XII

PRESSURE GAS-PRODUCERS

CHAPTER XIII

SUCTION GAS-PRODUCERS

CHAPTER I

MOTIVE POWER—COST OF INSTALLATION

THE ease with which a gas-engine can be installed, compared with a steam-engine is self-evident. In places where illuminating gas can be obtained and where less than 10 to 15 horse-power is needed, street-gas is ordinarily employed.* The improvements which have very recently been made in the construction of suction gas-generators, however, would seem to augur well for their general introduction in the near future, even for very small powers.

The installation of small street-gas-engines involves simply the making of the necessary connections with gas main and the mounting of the engine on a small base.

An economical steam-engine of equal power would necessitate the installation of a boiler and its setting, the construction of a smoke-stack, and other accessories, while the engine itself would require a firm base. Without exaggeration it may be asserted that the installation of a steam-engine and of its boiler requires five times as much time and trouble as the installation of a gas-engine of equal power, without considering even the requirements imposed by storing the fuel (Fig. 1). Small steam-engines mounted on

* Recent improvements made in suction gas-producers will probably lead to the wide introduction of producer gas engines even for small power.

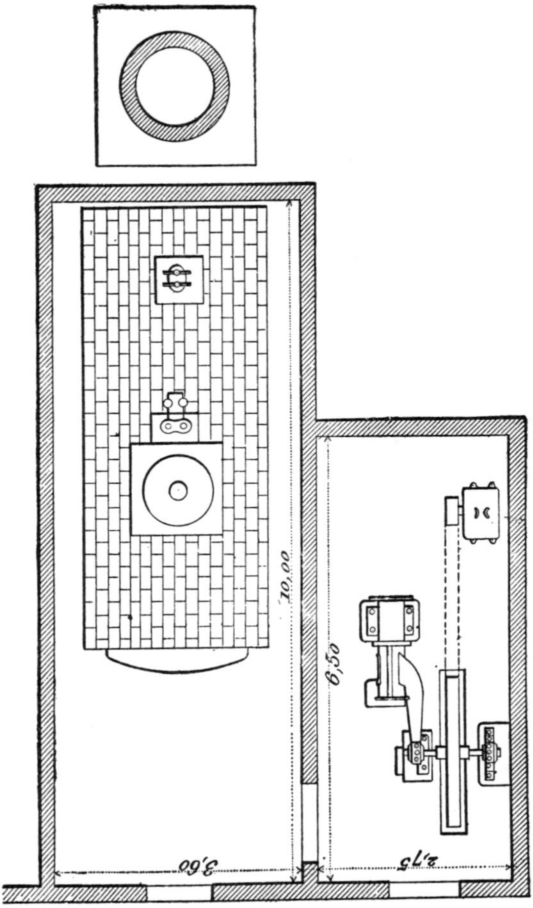

FIG. 1.
30 H. P. Steam-engine, boiler and smokestack.

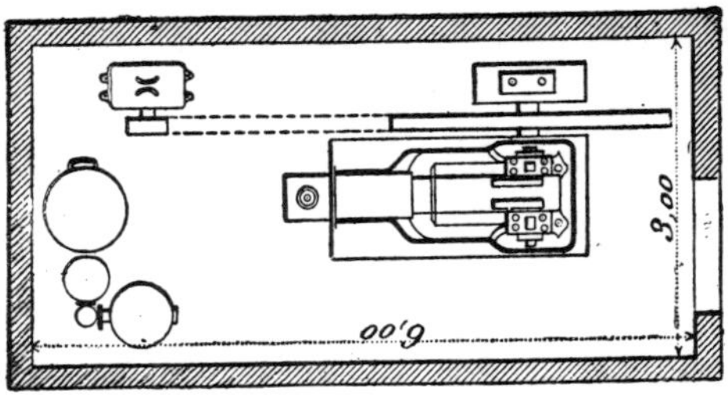

30 H. P. Gas-engine and suction gas-producer.

their own boilers, or portable engines, the consumption of which is generally not economical, are not here taken into account.

So far as the question of cost is concerned, we find that a 15 to 20 horse-power steam-engine working at a pressure of 90 pounds and having a speed of 60 revolutions per minute would cost about 16⅔ per cent. more than a 15 horse-power gas-engine, with its anti-pulsators and other accessories. The foundation of the steam-engine would likewise cost about 16⅔ per cent. more than that of the gas-engine. Furthermore the installation of the steam-engine would mean the buying of piping, of a boiler of 100 pounds pressure, and of firebrick, and the erection of a smoke-stack having a height of at least 65 feet. Beyond a little excavating for the engine-base and the necessary piping, a gas-engine imposes no additional burdens. It may be safely accepted that the steam-engine of the power indicated would cost approximately 45 per cent. more than the gas-engine of corresponding power.

The cost of running a 15 to 20 horse-power steam-engine is likewise considerably greater than that of running a gas-engine of the same size. Considering the fuel-consumption, the cost of the lubricating oil employed, the interest on the capital invested, the cost of maintenance and repair, and the salary of an engineer, it will be found that the operation of the steam-engine is more expensive by about 23 per cent.

This economical advantage of the gas over the steam-engine holds good for higher power as well, and becomes even more marked when producer-gas is used instead of street-gas. Comparing, for example, a 50 horse-power steam-engine having a pressure of 90 pounds and a speed of 60 revolutions per minute, with a 50 horse-power producer-gas engine, and considering in the case of the steam-engine the cost of a boiler of suitable size, foundation, firebrick, smoke-stack, etc., and in the case of the gas-engine the cost of the producer, foundation, and the like, it will be found that the installation of a steam-engine entails an expenditure 15 per cent. greater than in the case of the producer-gas engine. However, the cost of operating and maintaining the steam-engine of 50 horse-power will be 40 per cent. greater than the operation and maintenance of the producer-gas engine.

From the foregoing it follows that from 15 to 20 up to 500 horse-power the engine driven by producer-gas has considerably the advantage over the steam-engine in first cost and maintenance. For the development of horse-powers greater than 500, the employment of compound condensing-engines and engines driven by superheated steam considerably reduces the consumption, and the difference in the cost of running a steam- and gas-engine is not so marked. Still, in the present state of the art, superheated steam installations entail considerable expense for their maintenance and repair, thereby lessening their practical advantages and rendering their use rather burdensome.

CHAPTER II

THE SELECTION OF AN ENGINE

EXPLOSION-ENGINES are of many types. Gas-engines, of the four-cycle type, such as are industrially employed, will here be principally considered.

The Otto Cycle.—The term " four-cycle " motor, or Otto engine, has its origin in the manner in which the engine operates. A complete cycle comprises four distinct periods which are diagrammatically reproduced in the accompanying drawings.

The First Period.—*Suction:* The piston is driven forward, creating a vacuum in the cylinder, and simultaneously drawing in a certain quantity of air and gas (Fig. 2).

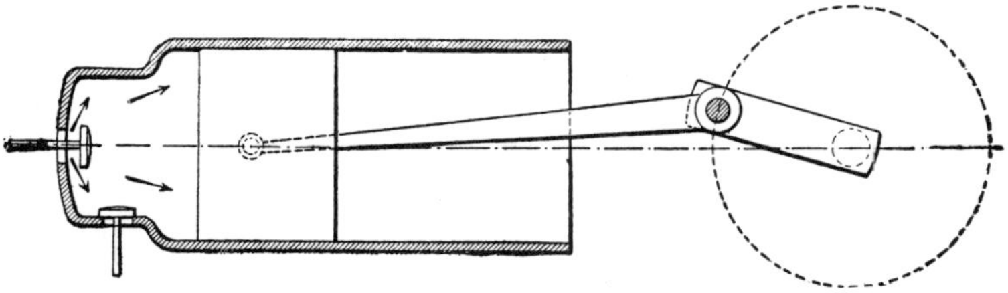

FIG. 2.—First cycle: Suction.

The Second Period.—*Compression:* The piston returns to its initial position. All admission and exhaust valves are closed (Fig. 3). The mixture drawn in during the first period is compressed.

The Third Period.—*Explosion and Expansion:* When the piston has reached the end of its return stroke, the compressed mixture is ignited. Explosion takes place at the dead center. The expansion of the gas drives the piston forward (Fig. 4).

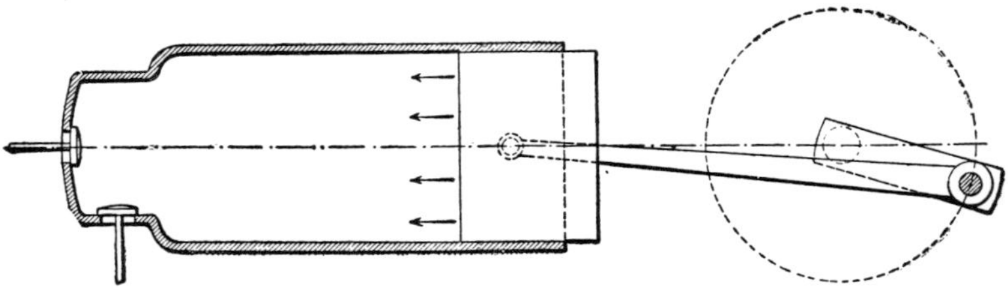

FIG. 3.—Second cycle: Compression.

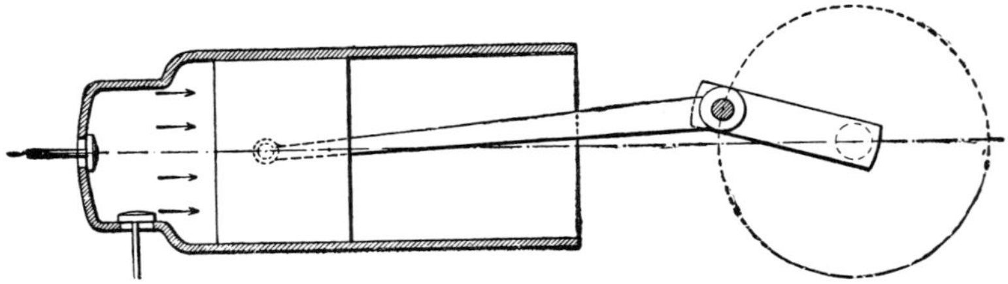

FIG. 4.—Third cycle: Explosion and expansion.

The Fourth Period.—*Exhaust:* The piston returns a second time. The exhaust-valve is opened, and the products of combustion are discharged (Fig. 5).

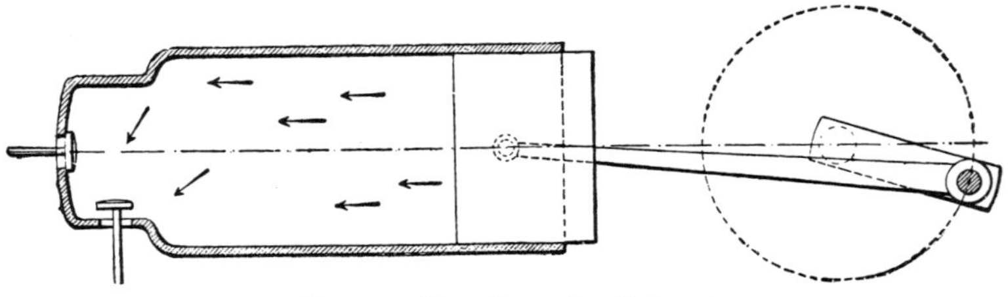

FIG. 5.—Fourth cycle: Exhaust.

These various cycles succeed one another, passing through the same phases in the same order.

Valve Mechanism.—It is to be noted that in modern motors valves are used which are better adapted to the peculiarities of explosion-engines than were the old slide-valves used when the Otto engine was first introduced. The slide-valve may now be considered as an antiquated distributing device with which it is impossible to obtain a low consumption.

In old-time gas-engines rather low compressions were used. Consequently a very low explosive power of the gaseous mixture, and low temperatures were obtained. The slide-valves were held to their seats by the pressure of external springs, and were generously lubricated. Under these conditions they operated regularly. Nowadays, the necessity of using gas-engines which are really economical has led to the use of high compressions with the result that powerful explosions and high temperatures are obtained. Under these conditions slide-valves would work poorly. They would not be sufficiently tight. To lubricate them would be difficult and ineffective. Furthermore, large engines are widely used in actual practice, and with these motors the frictional resistance of large slide-valves, moving on extensive surfaces would be considerable and would appreciably reduce the amount of useful work performed.

By reason of its peculiar operation, the slide-valve is objectionable, the gases being throttled at the time of their admission and discharge. As a result of these objections there are losses in the charge; and obnoxious counter-pressures occur. The necessity of using elements simple in their operation and free from the objections which have been mentioned, has naturally led to the adoption of the present valve. This valve is used both for the suction of the gas and of the air, as well as for the exhaust, with the result that either of these two essential phases in the operation of the motor can be independently controlled. The valves offer the following advantages: Their tightness increases with the pressure, since they always open toward the interior of the cylinder (Fig. 6). They have no rubbing surfaces,

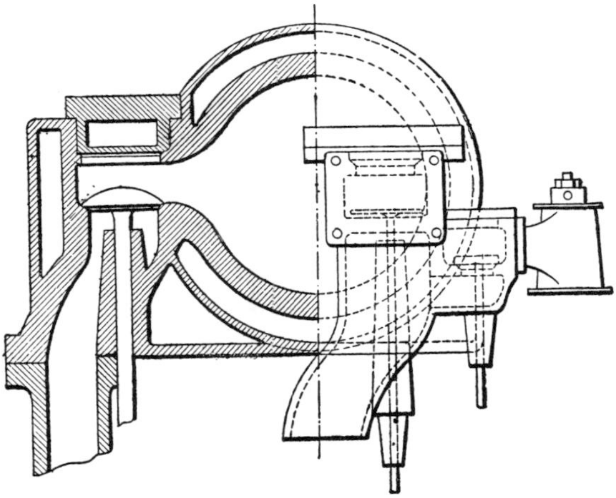

FIG. 6.—Modern valve mechanism.

and need not, therefore, be lubricated. Their opening is controlled by levers provided with quick-acting cams; and their closure is effected by coiled springs almost instantaneous in their action (Fig. 7). Each valve, depending upon the purpose for which it is used, can be mounted in that part of the cylinder best suited for its particular function. The types of valved motors now used are many and various. In order to attain economy in consumption and regularity in operation they should meet certain essential requirements which will here be reviewed.

Apart from proportioning the areas properly and from providing a suitable means of operation, it is indispensable that the valves should be readily accessible. Indeed, the valves should be regularly examined,

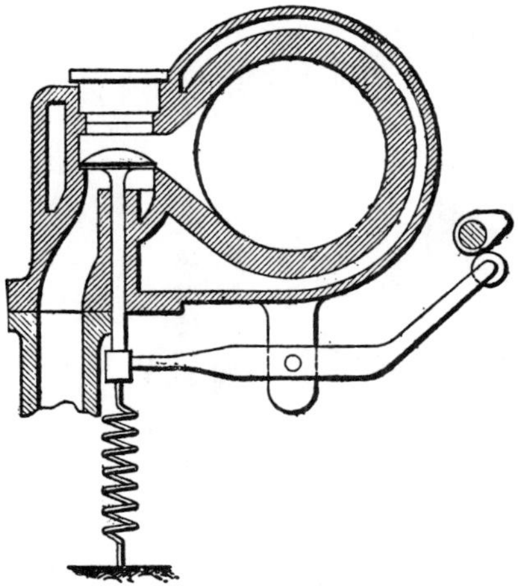

FIG. 7.—Controlling mechanism of valve.

cleaned and ground. It follows that it should be possible to take them apart easily and quickly.

It is necessary that the exhaust-valve be well cooled; otherwise the valve, exposed as it is to high temperatures, will suffer derangement and may cause leakage. The water-jacket should, therefore, surround the seat of the exhaust-valve, care being taken that the cooling water be admitted as near to it as possible (Fig. 8). The motor should control the air-let valve or that of the gaseous mixture. Hence these valves should not be actuated simply by springs, because springs are apt to move under the influence of the vacuum produced by suction.

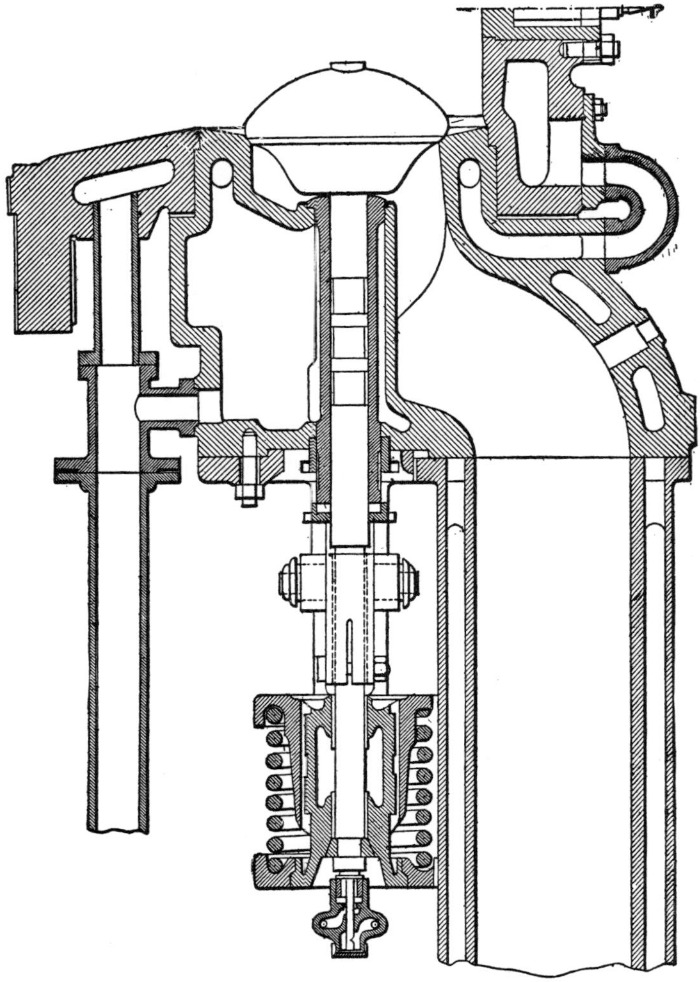

FIG. 8.—Water-jacketed valve.

The mixture of gas and air should not be admitted into the cylinder at too low a pressure; otherwise the weight of the mixture admitted would be lower than it ought to be, inasmuch as under these conditions the valve will be opened too tardily and closed prematurely. At the beginning as well as at the end of its stroke the linear velocity of the piston is quite inadequate to create a vacuum sufficient to overcome the resistance of the spring. It is, therefore, generally the practice separately to control the opening or closing of the one or the other valve (gas-valve or mixture-valve). Consequently these valves must be actuated independently of each other. Nowadays they are mechanically controlled almost exclusively,—a method which is advocated by well-known designers for industrial motors in particular. Valves which are not actuated in this manner (free valves) have only the advantage of simplicity of operation. Nevertheless, this arrangement is still to be found in certain oil and benzine engines, notably in automobile-motors. In these motors it is necessary to atomize the liquid fuel by means of aspired air, in order to produce an explosive, gaseous mixture.

Ignition.—In the development of the gas-engine, the incandescent tube and the electric spark have taken the place of the obsolete naked flame. The last-mentioned mode of exploding the gaseous mixture will not, therefore, be discussed.

The hot tube of porcelain or of metal has the indisputable merit of regularity of operation. The methods by which this operation is made as perfect as possible are many. Since certainty of ignition is obtained by means of the tube, it is important to time the ignition, so that it shall occur exactly at the moment when the piston is at the dead center. It has been previously stated that premature or belated ignition of the explosive mixture appreciably lessens the amount of useful work performed by the expansion of the gas. If ignition occur too soon, the mixture will be exploded before the piston has reached the dead center on its return stroke. As a result, the piston must overcome a considerable resistance due to the premature explosion and the consequent pressure. Furthermore, by reason of the high temperature of explosion, the gaseous products are very rapidly cooled. This rapid cooling causes a sudden drop in the pressure; and since a certain interval elapses between the moment of explosion and the moment when the piston starts on its forward stroke, the useful motive effort is the more diminished as the ignition is more premature.

Incandescent Tubes.—In Figs. 9 and 10 two systems most commonly used are illustrated. In these two arrangements, in which no valve is used, the length or height to which the tube is heated by the outer flame is so controlled that the gaseous mixture, which has been driven into the tube after compression, reaches the incandescent zone as nearly as possible at the exact moment when ignition and explosion should take place. The temperature of the flame of the burner, the richness of the gaseous mixture, and other circumstances, however, have a marked influence on the time of ignition, so that the mixture is never fired at the exact moment mentioned.

These considerations lead to the conclusion that

motors in which the mixture is exploded by hot tubes provided with an ignition-valve are preferable to valveless tubes. By the use of a special valve, positively controlled by the motor itself, the chances of untimely ignition are lessened, because it is necessary simply to regulate the temperature and the position of the tube in order that ignition may be surely effected imme-

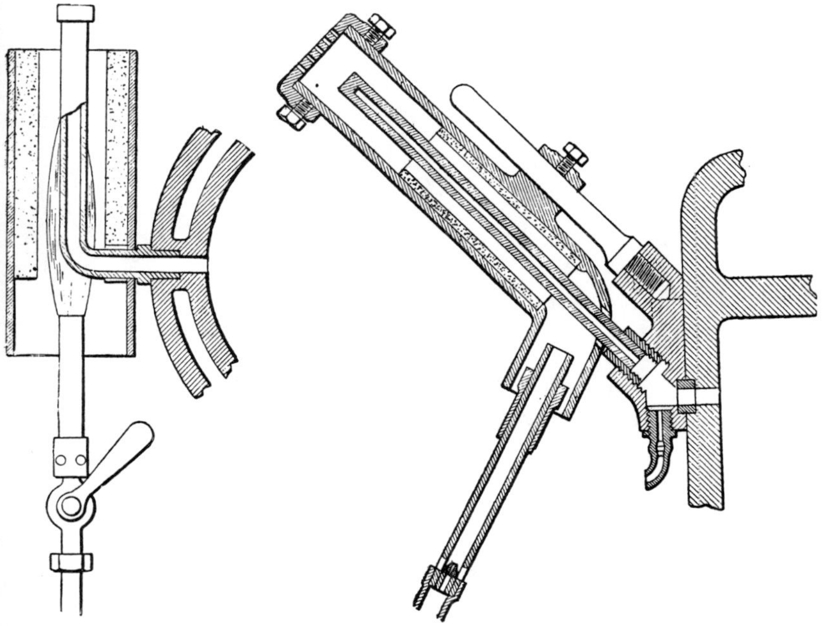

FIGS. 9–10.—Valveless hot tubes.

diately upon the opening of the valve, at the very moment the cylinder gases come into contact with the incandescent portion of the tube (Fig. 11). Many manufacturers, however, do not employ the ignition-valve on motors of less than 15 to 20 horse-power, chiefly because of the cheaper construction. The total consumption is of less moment in a motor of small than of great power, and the loss due to the lack of an ignition-valve not so marked. In a high-power engine, premature explosion may be the cause of the breaking of a vital part, such as the piston-rod or the crank-shaft. For this reason, a valve is indispensable for engines of more than 2 oto 25 horse-power. A breakage of this kind is less to be feared in a small motor, where the parts are

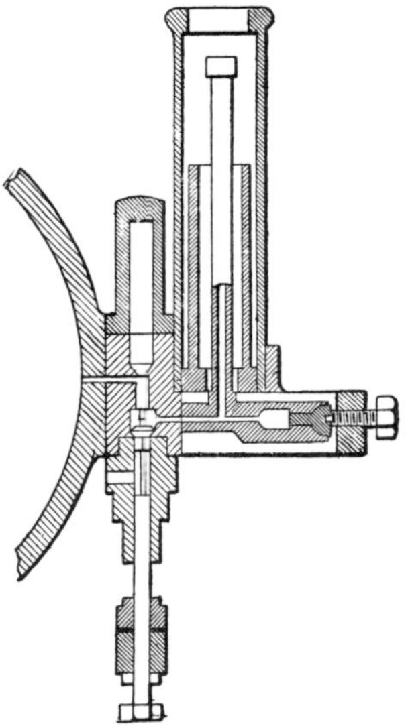

FIG. 11.—Ignition-tube with valve.

comparatively stout. The gas consumption of a well-designed burner does not exceed from 3.5 to 5 cubic feet per hour.

Electric Ignition.—Electric ignition consists in producing a spark in the explosion-chamber of the engine. The nicety with which it can be controlled gives it an undeniable advantage over the hot tube. But the objection has been raised, perhaps with some force, that it entails certain complications in installing the engine. Its opponents even assert that the power and the rapidity of the deflagration of the explosive mixture are greater with hot-tube ignition. This reason may have caused the hot-tube system to prevail in England, where manufacturers of gas-engines are very numerous and not lacking in experience.

Electric ignition is effected in gas-engines by means of a battery and spark-coil, or by means of a small magneto machine which mechanically produces a current-breaking spark.

Electric Ignition by Battery and Induction-Coil.—The first system is the cheaper; but it exacts the most

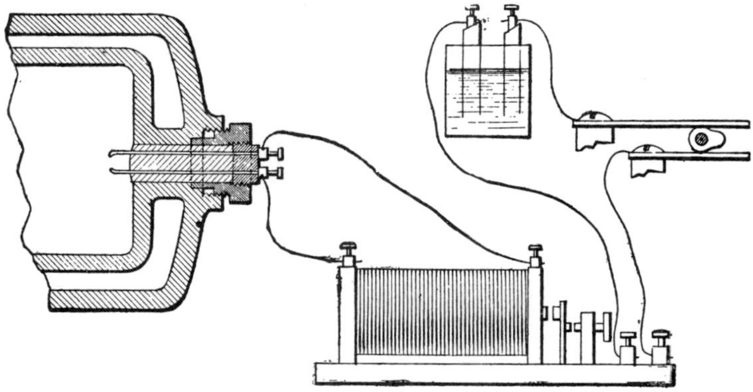

FIG. 12.—Electric ignition by spark-coil and battery.

painstaking care in maintaining the parts in proper working condition. It comprises three essential elements—a battery, a coil, and a spark-plug (Fig. 12). The battery may be a storage-battery, which must, consequently, be recharged from time to time; or it may be a primary battery which must be frequently renewed and carefully cleaned. The induction-coil is fitted with a trembler or interrupter, which easily gets out of order and which must be regulated with considerable accuracy. The spark-plug is a particularly delicate part, subject to many possible accidents. The porcelain of which it is made is liable to crack. It is hard to obtain

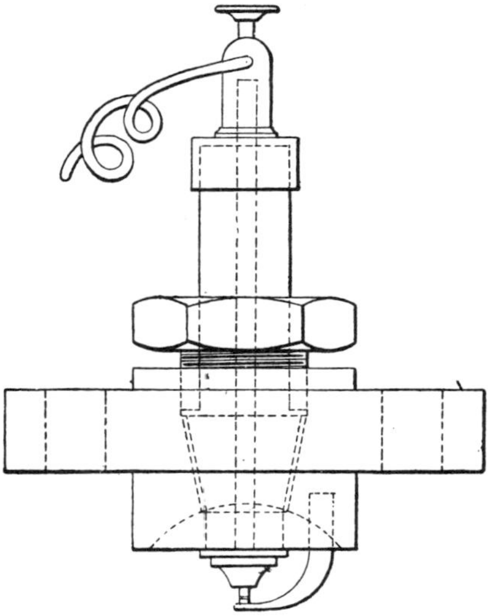

FIG. 13.—Spark-plug.

absolutely perfect insulation; for the terminals deteriorate as they become overheated, break, or become foul (Fig. 13). In oil-engines, especially, soot is rapidly deposited on the terminals, so that no spark can be produced. In benzine or naphtha motors, such an accident is less likely to happen. In automobile-motors, however, the spark-plug only too often fails to perform its function. The one remedy for these evils is to be found in the most painstaking care of the spark-plug and of the other elements of the ignition system.

Ignition by Magnetos.—Magneto apparatus, on the other hand, are noteworthy for the regularity of their operation. They may be used for several years with-

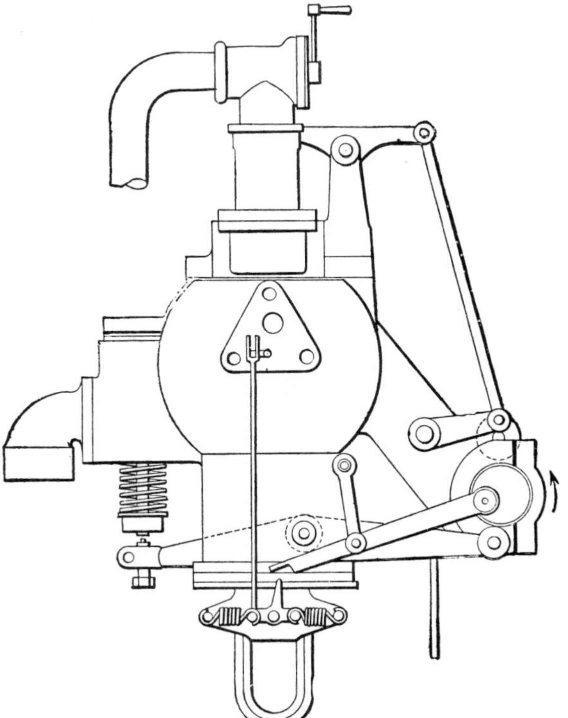

FIG. 14.—Magneto ignition apparatus.

out being remagnetized, and require no exceptional care. Magneto ignition devices are mechanically actuated, the necessary displacement of the coil being effected by means of a cam carried on a shaft turning with half the motor speed (Figs. 14 and 15). At the moment when it is released by the cam, the coil is suddenly returned to its initial position by means of a spring. This rapid movement generates a current that passes through terminals, which are arranged within the cylinder and which are immediately separated by mechanical means. Thus a much hotter circuit-break-

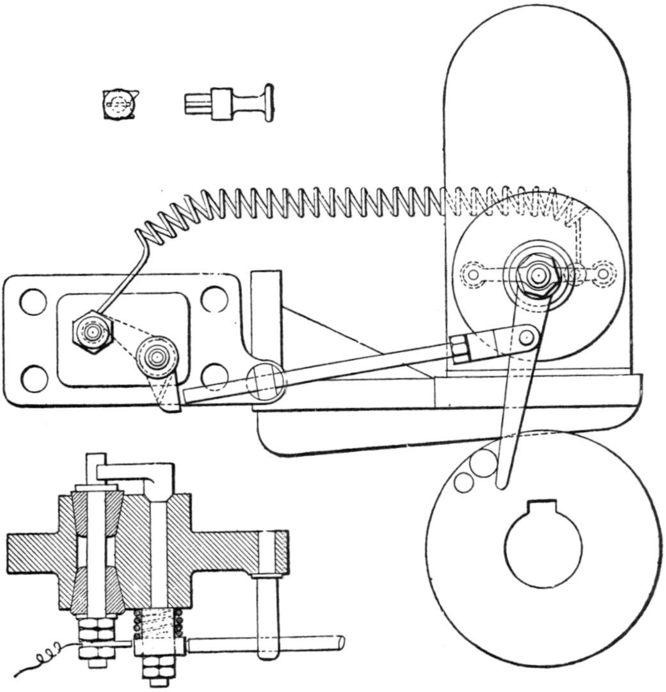

FIG. 15.—General view and details of a magneto ignition apparatus.

ing spark is produced, which is very much more energetic than that of a battery and induction-coil, and which surely ignites the gaseous mixture in the cylinder. The terminals are generally of steel, sometimes pointed with nickel or platinum (Fig. 16). The only precaution to be observed is the exclusion of moisture and occasional cleaning. For engines driven by producer-gas magneto-igniters are preferable to cells and

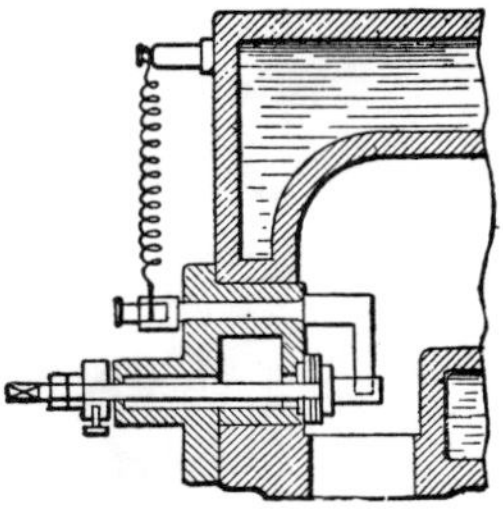

FIG. 16.—Contacts of a magneto-igniter.

batteries. In general, electrical ignition is to be recommended for high-pressure engines.

In order to explain more clearly modern methods of ignition a diagram is presented, showing an electric

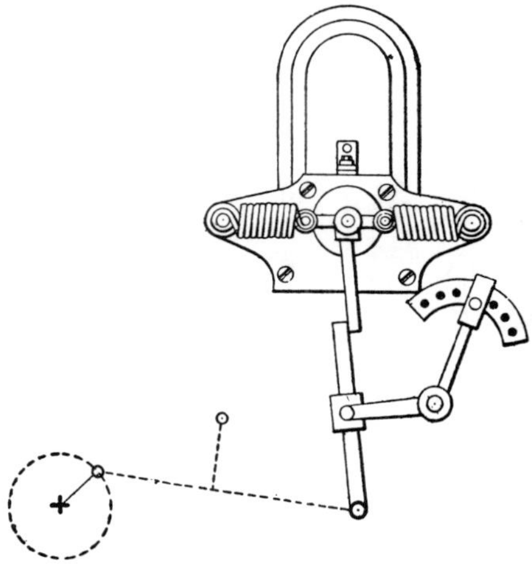

FIG. 17.—Device for regulating the moment of ignition.

magneto-igniter applied to the cylinder-head of a Winterthur motor, and also a sectional view of the member varying the make-and-break contacts which are mounted in the explosion-chamber (Figs. 18 and 19)

1. The magneto *A* consists of horseshoe-magnets, between the poles of which the armature rotates. At its conically turned end, the armature-shaft carries an arm *B*, held in place by a nut.

FIG. 18.—Winterthur electric ignition system.

2. The igniter *C* is a casting secured to the cylinder-head by a movable strap and provided with two axes *D* and *M*, of which the one, *D*, made of bronze, is movable, and is fitted with a small interior contact-hammer, a percussion-lever, and an exterior recoil-spring; the other, *M*, is fixed, insulated, and arranged

to receive the current from the magneto *A*, by means of an insulated copper wire *E*.

3. The spring *F* comprises two continuous coils contained in a brass casing, and actuating a steel striking or percussion-pin.

4. The controlling devices of the magneto include a stem or rod *G*, slidable in a guide *H*, provided with a

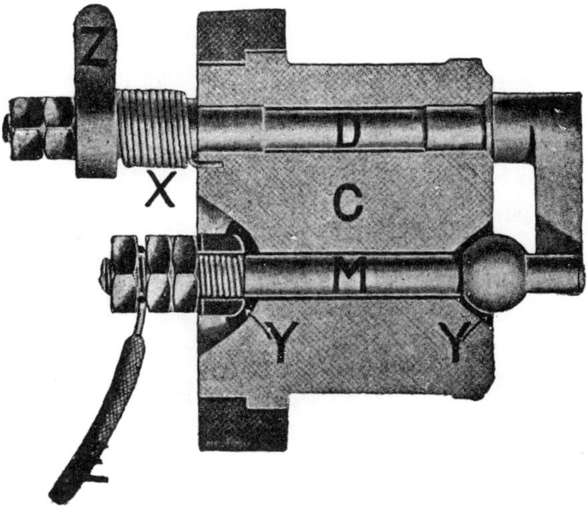

FIG. 19.—Contacts of the Winterthur system.

safety spring and mounted on an eccentric spindle, the position of which can be varied by means of a regulating-lever (*I*). The rod is operated from the distributing-shaft, on the conical end of which a cam *J* carrying a spindle is secured.

Regulation of the Magneto.—The position assumed by the armature when at rest is a matter of importance in obtaining a good spark on breaking the circuit. The marks on the armature should be noted. The position of the armature may be experimentally varied, in order to obtain a spark of maximum intensity, by changing the position of the arm *B* on the armature-shaft.

Control of the Magneto.—The controlling gear should enable the armature to oscillate from 20 to 25 degrees. The time at which the breaking of the circuit is effected can be regulated by shifting the handle (*I*). In starting the engine, the circuit can be broken with a slight retardation, which is lessened as the engine attains its normal speed.

Igniter.—It is advisable that there should be a play of ½ mm. (0.0196 in.) between the lever *Z* when at rest and the striking-pin. The axis *D* of the circuit-breaking device should be easily movable; and the hammer which it carries at its end toward the interior of the cylinder should be in perfect contact with the stationary spindle *M*, which is electrically insulated. This spindle *M* should be well enclosed, in order to prevent any leakage that might cause a deterioration of the insulating material.

The subject of ignition is of such extreme importance that the author will recur to it from time to time in the various chapters of this book. Too much stress cannot be laid upon proper timing; otherwise there will be a needless waste of power. Cleanliness is a point that must be observed scrupulously; for spark-plugs are apt to foul only too readily, with the result that short-circuits and misfires are apt to occur. In oil and volatile hydrocarbon engines the tendency to fouling is particularly noticeable. In the chapter devoted to these forms of motors the author has dwelt upon the precautions that should be taken to forestall a possible derangement of the ignition apparatus. As a general rule the ignition apparatus installed by trustworthy manufacturers will be found best suited for the requirements of the engine.

The apparatus should be fitted with a device by which the ignition can be duly timed by hand during operation (Fig. 17).

The Piston.—Coming, as it does, continually in contact with the ignited gases, the piston is gradually

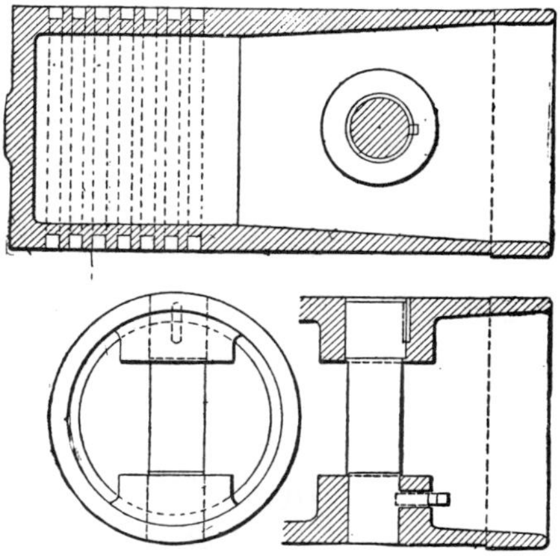

FIG. 20.—Design of the piston.

heated to a high temperature. The rear face of the piston should preferably be plane. Curved surfaces are not to be recommended because they cool off badly. Likewise, faces having either inserted parts or bolt-heads are to be avoided, since they are liable to become red-hot and to ignite the mixture prematurely (Fig. 20).

Among the parts of the piston which rapidly wear away because constant lubrication is difficult, is the connection with the piston-rod (Fig. 21). It is important that the bearing at the piston-pin be formed of two parts which can be adjusted to take up the wear. The pin itself should be of case-hardened steel. For large engines, some manufacturers have apparently abandoned the practice of locking the pin, by set-screws, in flanges cast in one piece with the piston. Indeed, the piston is often fractured by reason of the expansion of the pins thus held on two sides. It seems advisable

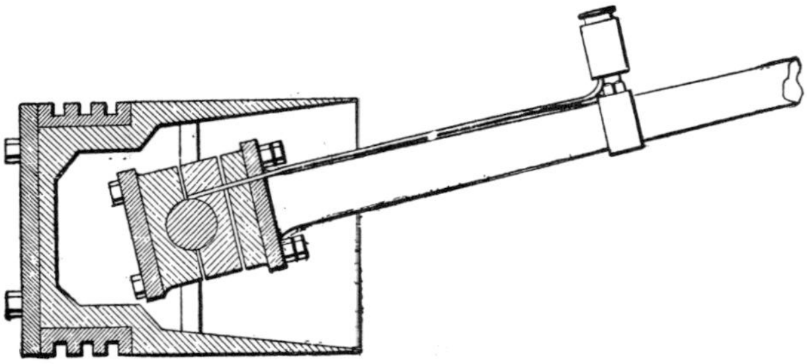

FIG. 21.—Piston with lubricated pin.

to secure the pin by means of a single screw in one of the flanges, fitting it by pressure against the opposite boss. The use of wedges or of clamping-screws, introduced from without the piston to hold the pin, should be avoided. It may happen that the wedges will be loosened, will move out, and will grind the cylinder, causing injuries that cannot be detected before it is too late. The strength of the piston-pin should be so calculated that the pressure per square inch of projected surface does not exceed 1,500 to 2,850 pounds per square inch. It should be borne in mind that the initial pressure of the explosion is often equal to 400 to 425 pounds per square inch. Some manufacturers mount the pin as far to the back of the piston as possible, so as to bring it nearer the point of application of the motive force of the explosion. Other manufacturers, on the other hand, mount the pin toward the front of the piston. No great objection can be raised against either method. In the former case the position of the rings will limit that of the pin.

The number of these rings ought not to be less than four or five, arranged at the rear of the piston. It is to be observed that makers of good engines use as many as 8 to 10 rings in the pistons of fair-sized motors.

Piston-rings of gray pig-iron can be adjusted with the greatest nicety in such a manner that, by means of tongues fitting in their grooves, they are held from turning in the latter, whereby their openings are prevented from registering and allowing the passage of gas. As a general rule, a large number of rings may be considered a distinguishing feature of a well-built engine. In order to prevent a too rapid wear of the cylinder, several German manufacturers finish off the front of the piston with bronze or anti-friction metal in engines of more than 40 to 50 horse-power. It is to be observed, however, that this expedient is not applicable to motors the cylinders of which are comparatively cold; otherwise the bronze or anti-friction metal will deteriorate.

Arrangement of the Cylinder.—The cylinder shell or liner, in which the piston travels, and the water-jacket should preferably be made in separate pieces and not cast of the same metal, in order to permit a free expansion (Figs. 22 and 23). If for want of care or

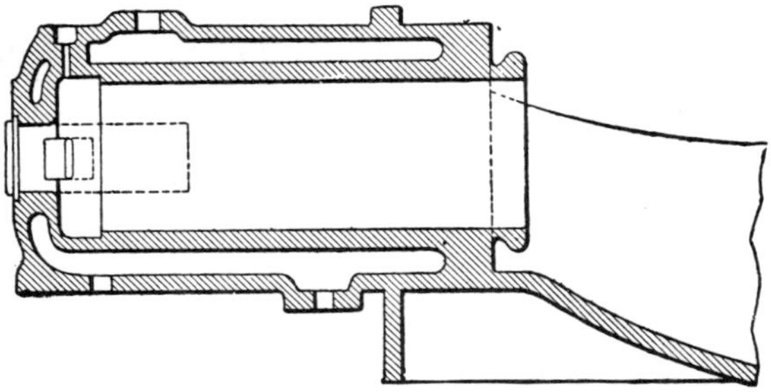

FIG. 22.—Head, jacket and liner of cylinder, 'cast in one piece.

of proper lubrication, which frequently occurs in gas-engines, the cylinder should be injured by grinding, it can be easily renewed, without the loss of all the connecting parts.

For the same reason, the cylinder and its casing

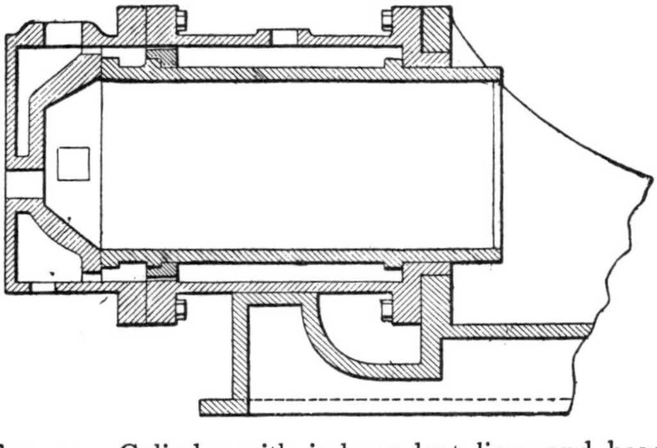

FIG. 23.—Cylinder with independent liner and head.

should be independent of the frame. In many horizontal engines, the cylinders overhang the frame throughout the entire length, by reason of the joining of their front portions with the frames. Although such a construction is attended with no serious consequences in small engines, nevertheless in large engines it is exceedingly harmful. Indeed, in most modern single-acting engines, the pistons are directly connected with the crank-shaft by the piston-rod, without any intermediate connecting-rod or cross-head. The vertical re-

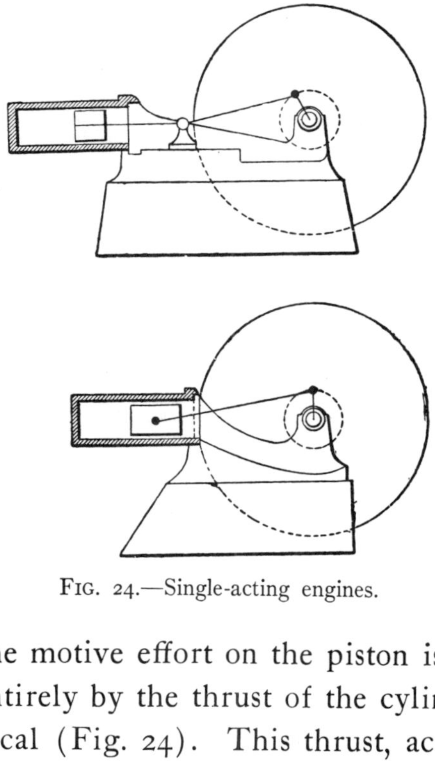

FIG. 24.—Single-acting engines.

action of the motive effort on the piston is, therefore, taken up entirely by the thrust of the cylinder, which is also vertical (Fig. 24). This thrust, acting against an unsupported part, may cause fractures; at any rate, it entails a rapid deterioration of the cylinder joint.

The Frame.—Gas-engines driven as they are, by explosions, giving rise to shocks and blows, should be built with frames, heavy, substantial, and broad-based, so as to rest solidly on the ground. This essential condition is often fulfilled at the cost of the engine's appearance; but appearance will be willingly sacrificed to meet one of the requirements of perfect operation. For engines of more than 8 to 10 horse-power, frames should be employed which can be secured to the masonry foundation without a separate pedestal or base. Some manufacturers, for the purpose of lightening the frame, attach but little importance to the foundation

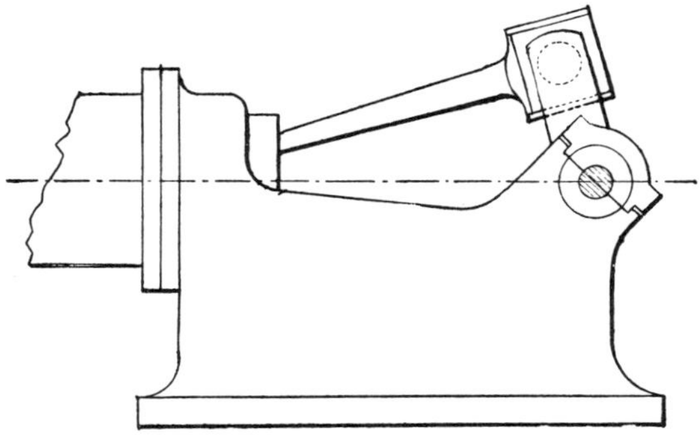

FIG. 25.—Engine with inclined bearings.

and to strength of construction, and employ the design illustrated in place of the crank-shaft bearing (Fig. 25); others, in order to facilitate the adjusting of the connecting-rod bearings, prefer the second form (Fig. 26). It is evident that, in the first case, a part of the effort produced by the explosion reacts on the upper portion of the connecting-rod bearing, on the cap of the crank-shaft bearing, and consequently on the fastening-bolts. In the second case, if the adjustment be not very carefully made, or if the rubbing surfaces are insufficient, the entire thrust due to the explosion will be received by the meeting parts of the two bushings, thus injuring them and causing a more rapid wear. In the construction of large engines, some manu-

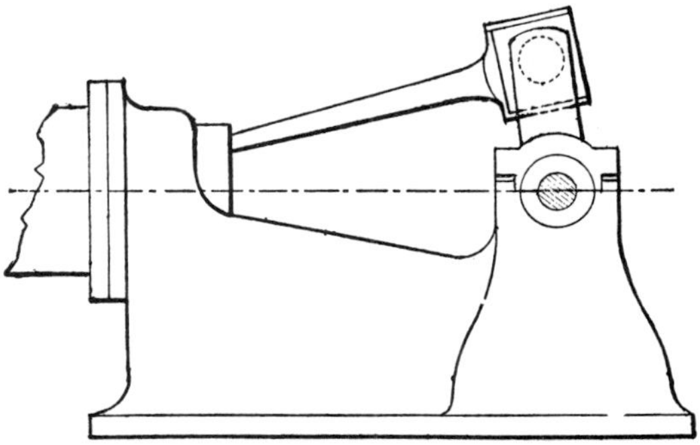

FIG. 26.—Engine with straight bearings.

facturers take the precaution of forming the connecting-rod bearings of four parts, adjustable to take up the wear, so that the effort is exerted against the parts disposed at right angles to each other. A form that seems rational is that shown in Fig. 27, in which the reaction

FIG. 27.—Engine with correctly designed bearings.

of the thrust is taken up by the lower bearing, rigidly supported by the braced frame, in the direction opposite to that of the explosive effort.

The sum of the projecting surfaces of the two bearings should be so calculated that a maximum explosive pressure of 405 to 425 pounds per square inch will not subject the bearings to a pressure higher than 425 to 550 pounds per square inch.

Fly-Wheels.—In gas-engines particularly, the fly-wheel should be secured to the crank-shaft with the utmost care. It should be mounted as near as possible to the bearings; otherwise the alinement of the shaft will be destroyed and its strength impaired. If the fly-wheel be fastened by means of a key or wedge having a projecting head, it is advisable to cover the end of the shaft by a movable sleeve. The fly-wheel should run absolutely true and straight even if the explosion be premature. In well-built engines the fly-wheels are lined up and shaped to the rim. The periphery is slightly rounded in order the better to guide the belt when applied to the wheel.

Furthermore, fly-wheels should be nicely balanced; those are to be preferred which have no counter-weights cast or fastened to the hub, the spokes, or the rim.

The system of balancing the engine by means of two fly-wheels, mounted on opposite sides, is used chiefly for the purpose of equalizing the inertia effects. Special engines, employed for driving dynamos, and even industrial engines of high power, are preferably fitted with but a single fly-wheel, with an outer bearing, since they more readily counteract the cyclic irregularities or variations of speed occurring in a single revolution (Fig. 28). If in this case a pulley be provided, it

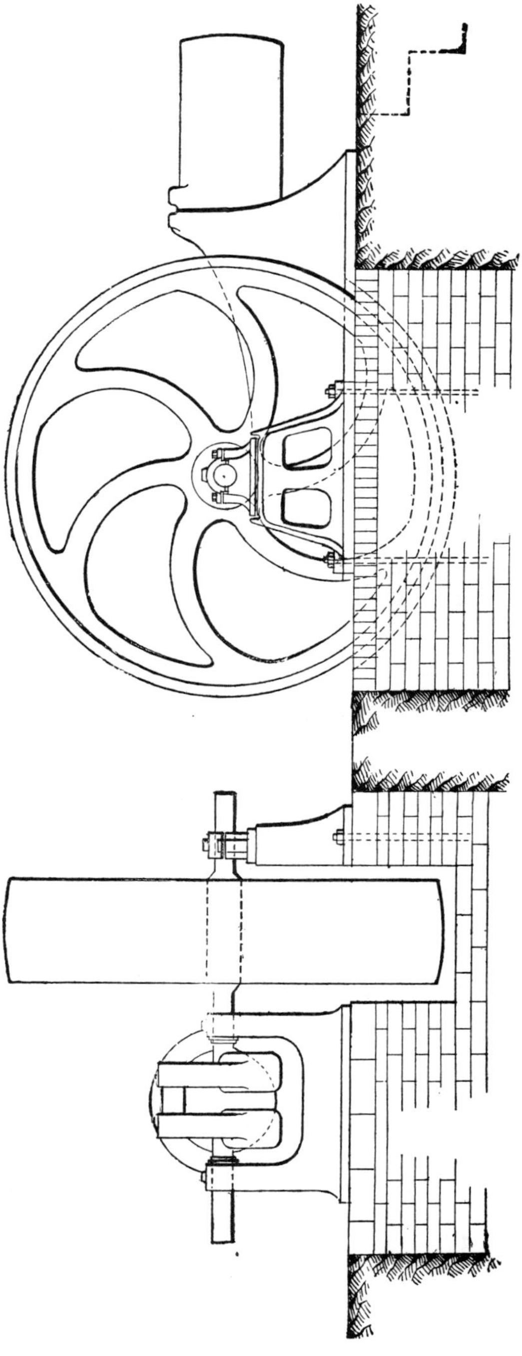

FIG. 28.—Single fly-wheel engine with external bearing.

should be mounted between the engine and the outer bearing. The following advantages may be cited in favor of the single fly-wheel, particularly in the case of dynamo-driving engines:

1. The single fly-wheel permits a more ready access to the parts to be examined.

2. It involves the employment of a third bearing, thus avoiding the overhang caused by two ordinary fly-wheels.

3. It avoids the torsional strain to which the two-wheel crank is subjected when starting, stopping, and changing the load, the peripheral resistance varying in one of the fly-wheels, while the other is subjected to a strain in the opposite direction on account of the inertia.

4. Two fly-wheels, keyed as they are to projecting ends of the shaft, will be so affected at the rims by the explosions that the belts will shake.

The third bearing which characterizes the single-fly-wheel system, is but an independent support, resting solidly on the masonry bed of the engine. The bearing with its independent support is sufficiently rigid, and is not subjected to any stress from the crank at the moment of explosion, the reaction of the crank affecting only the frame bearings. With such fly-wheels, reputable firms guarantee a cyclic regularity which compares favorably with that of the best steam-engines. For a duty varying from a third of the load to the maximum load, these engines, when driving direct-current dynamos for directly supplying an electric-light circuit, will insure perfect steadiness of the light; and the effectually aperiodic measuring instruments will not indicate fluctuations greater than 2 to 3 per cent. of the tension or intensity of the current. The coefficient of the variations in the speed of a single revolution will thus be not far from $\frac{1}{60}$.

Straight and Curved Spoke Fly-Wheels.—The

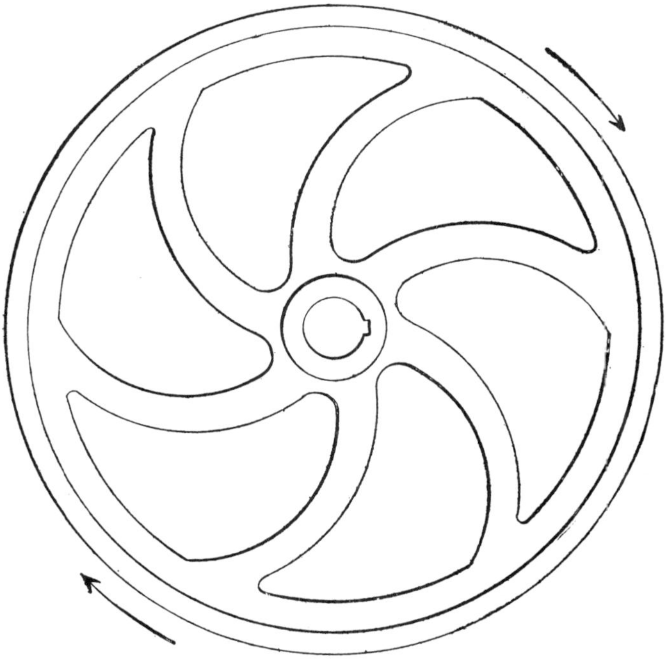

FIG. 29.—Curved spoke fly-wheel.

spokes of fly-wheels are either straight or curved. In assembling the motor parts it too often occurs that curved spoke fly-wheels are mounted with utter disregard of the direction in which they are to turn. It is important that curved spokes should be subjected to compression and not to traction. Hence the fly-wheels should be so mounted that the concave portions of the spokes travel in the direction of rotation, as shown in the accompanying diagram (Fig. 29). If a single fly-wheel be employed on an engine of the type in which

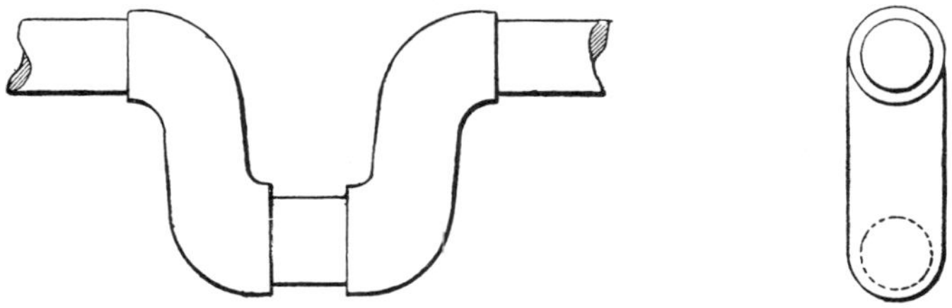

Fig. 30.—Forged crank-shafts.

the speed is governed by the "hit-and-miss" system, the fly-wheel should be extra heavy to counteract the irregularities of the motive impulses when the engine is not working at its full load, or in other words, when no explosion takes place at every cycle.

The Crank-Shaft.—The crank-shaft should be made of the best mild steel. Those shafts are to be preferred

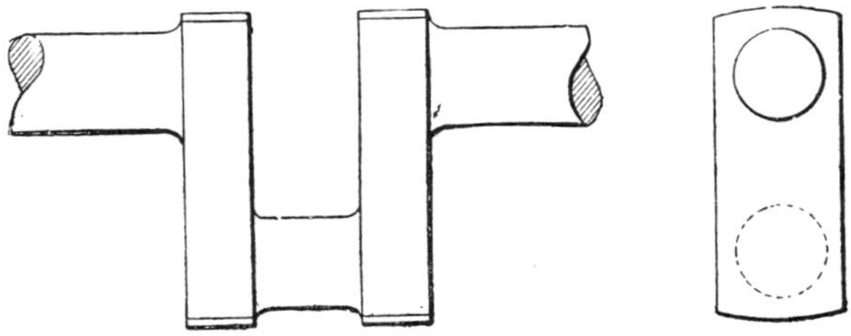

Fig. 31.—Correct design of crank-shaft.

the cranks of which are not forged on (Fig. 30), but cut out of the mass of metal; furthermore, the brackets or supports should be planed and shaped so that they are square in cross-section.

Such a design involves fine workmanship and speaks well for the construction of the whole engine. Moreover, it enables the bearings to be brought nearer each other, reduces to a minimum that part of the crank-shaft which may be considered the weakest, and permits a rational and exact counterbalancing of the moving parts, such as the crank and the end of the connecting-rod. The best manufacturers have adopted the method of fastening to the cranks balancing weights secured to the brackets, especially for high-speed en-

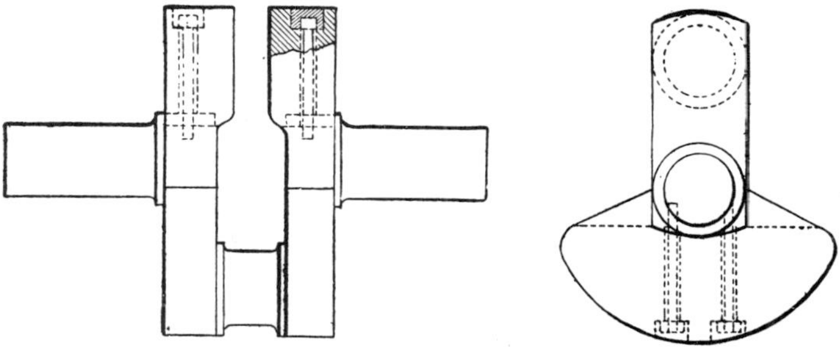

FIG. 32.—Crank-shaft with balancing weight.

gines or for engines of high power. The projecting surface of the crank-pin should, as a rule, be calculated for a pressure of 1,400 pounds per square inch.

Cams, Rollers, etc.—The cams, rollers, thrust-bearings, as well as the piston-pin in particular, should be made of good steel, case-hardened to a depth of at least .08 of an inch. Their hardness and the degree of cementation may be tested by means of a file. This is the method followed by the best manufacturers.

Bearings.—All the bearings and all guides should be adjustable to take up the wear. They are usually made of bronze or of the best anti-friction metal.

Steadiness.—The steadiness of engines may be considered from two different standpoints.

1. *Variation of the Number of Revolutions at Different Loads.*—This depends chiefly on the sensitiveness of the governor, which should be of the "inertia" or of the "ball" (or centrifugal) type. The first form is rarely employed, except in small engines up to 10-horse-power, and is applicable only to engines in which

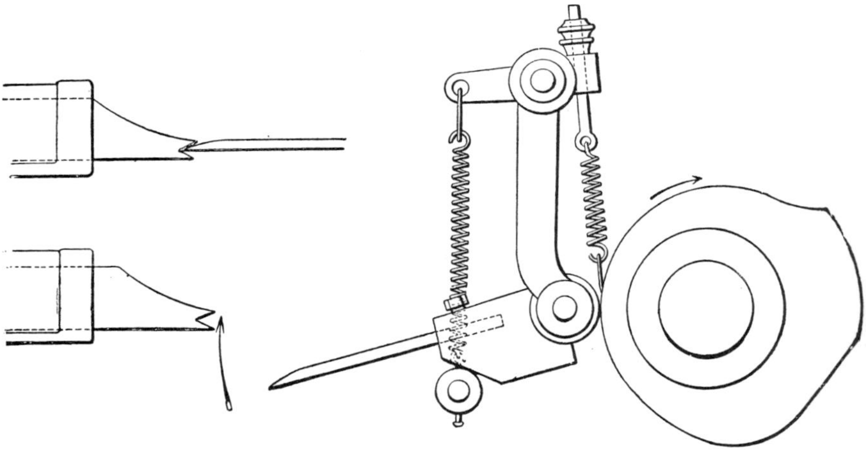

FIG. 33.—Inertia governor.

the "hit and miss" system is employed (Fig. 33). The second form is more widely used, and is applicable to engines having "hit-and-miss" or variable admission devices. In the first form, the governor simply displaces a very light member, whatever may be the size of the engine, for which reason the dimensions are very small. In the second form, on the other hand, the governor acts either on a conical sleeve or on some other regulating member offering resistance. Evidently, in order to overcome the reactions to which it is subjected, it must be as heavy and powerful as a steam-engine governor. Sufficient allowance is made in a good engine for variation in the number of revolutions between no load and full load, not greater than two per cent. if the admission be of the " hit-and-miss " type, and five per cent. if it be of the variable type.

2. *Cyclic Regularity.*—This term means simply that the speed of the engine is constant in a single revolution. In practice this is never attained. Allowance is made in engines used for driving direct-current dynamos for a variation of about $\frac{1}{60}$; while in industrial engines a variation of $\frac{1}{25}$ is permissible. Cyclic variation depends only on the weight of the fly-wheel; whereas variation in the number of revolutions is determined chiefly by the governor.

Governors.—Diagrams are here presented of the principal types of governors—the inertia governor, the ball or centrifugal governor controlling an admission-valve of the " hit-and-miss " type (Fig. 34), and the ball or centrifugal governor controlling a variable gas-admission valve (Fig. 35).

In distinguishing between the operation of the two last-mentioned types, it may be stated that the former bears the same relation to the hit-and-miss gear as it does, for example, to the valve gear of a Corliss steam-engine. In other words, it is an apparatus that *indicates* without *inducing,* admission or cut-off. The second type, on the other hand, operates by means of slides and the like, as in the Ridder type of engine, in which it controls the displacement of the cut-off or distribution slide-valve and is subjected to variable forces, depending on the pressure, lubrication, the condition of the stuffing-boxes, and the like.

In gas as well as in steam engines, designs are to be commended which shield the delicate mechanism from strains and stresses that are likely to destroy its sensi-

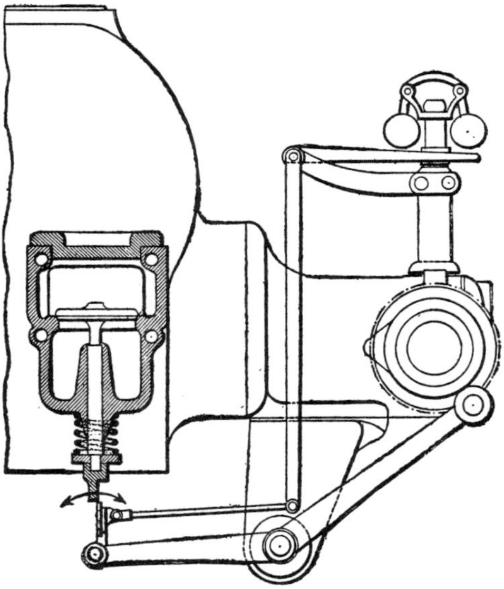

FIG. 34.—"Hit-and-miss" governor.

tiveness, as is the case in the automatic cut-off of the Corliss steam-engine.

Governors should be provided with means to permit the manual variation of the speed while the engine is in operation.

For small motors, one of the most widely used admission devices is that of the "hit-and-miss" type. As its name indicates, this admission arrangement allows a given quantity of gas to enter the cylinder for a number of consecutive intervals, until the engine is about to exceed its normal speed. Thereupon the governor cuts off the gas entirely. The result is that, in this system, the number of admissions is variable, but that each admitted charge is composed of a constant proportion of gas and air.

The governors employed for the "hit-and-miss" type are either "inertia" or "centrifugal" governors.

Inertia governors (Fig. 33) are less sensitive than those of the centrifugal type. They are generally applied only to industrial engines of small power, in which regularity of operation is a secondary consideration.

Centrifugal governors employed for gas-engines with "hit-and-miss" regulation are, as a general rule, noteworthy for their small size, which is accounted for by the fact that, in most systems, merely a movable member is placed between the admission-controlling means and the valve-stem (Fig. 34). It follows that this method of operation relieves the governor of the necessity of overcoming the resistance of the weight of moving parts, more or less effectually lubricated, and subjected to the reaction of the parts which they control.

In engines equipped with variable admission devices for the gas or the explosive mixture, the governor actuates a sleeve on which the admission-cam is fastened (Fig. 35). Or, the governor may displace a conical cam, the reaction of which, on contact with the lever, destroys the stability of the governor. These conditions justify the employment of powerful governors which, on account of the inertia of their parts, diminish the reactionary forces encountered.

The centrifugal governor should be sufficiently effectual to prevent variations in the number of revolutions within the limits of 2 to 3 per cent. between no load and approximately full load. Under equivalent

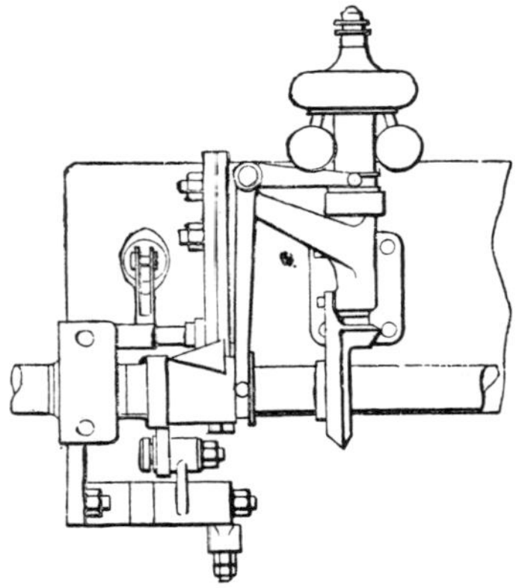

FIG. 35.—Variable admission governor.

conditions, the inertia governor can hardly be relied upon for a coefficient of regularity greater than 4 to 5 per cent.

The manner of a governor's operation is necessarily dependent on the admission system adopted. And the admission system varies essentially with the size, the purpose of the engine, and the character of the fuel employed.

Vertical Engines.—For some years past there seems to have been a tendency in Europe to use horizontal instead of vertical engines, especially since engines of more than 10 or 15 horse-power have been extensively used for industrial purposes. The vertical type is used for 1 to 8 horse-power engines, with the cylinder in the lower part of the frame, and the shaft and its fly-wheel in the upper part (Fig. 36). The only merit to be attributed to this arrangement is a great saving of space. It is evident, however, that beyond a certain size and power, such engines are unstable. In America particu-

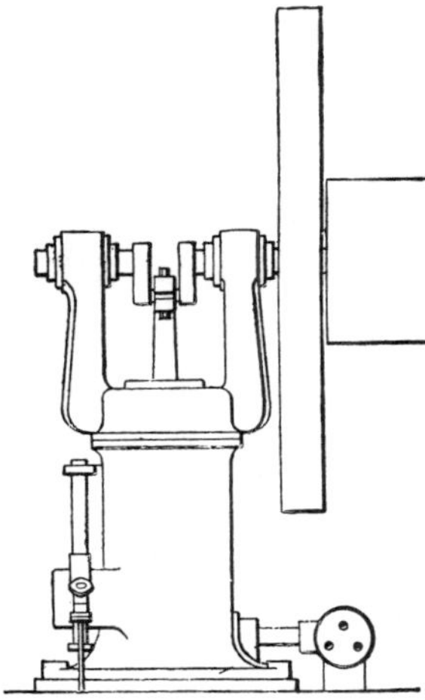

FIG. 36.—Vertical engine.

larly, many manufacturers of high-power engines (50 to 100 horse-power or more) prefer the vertical or "steam-hammer" arrangement, which consists in placing the cylinder in the upper part, and the shaft in the lower part of the frame as close to the ground as possible (Figs. 37 and 38). The problem of saving space, as well as that of insuring stability, is thus solved, so that it is easily possible to run up the speed of the engine. There is also the advantage that the shaft of a dynamo

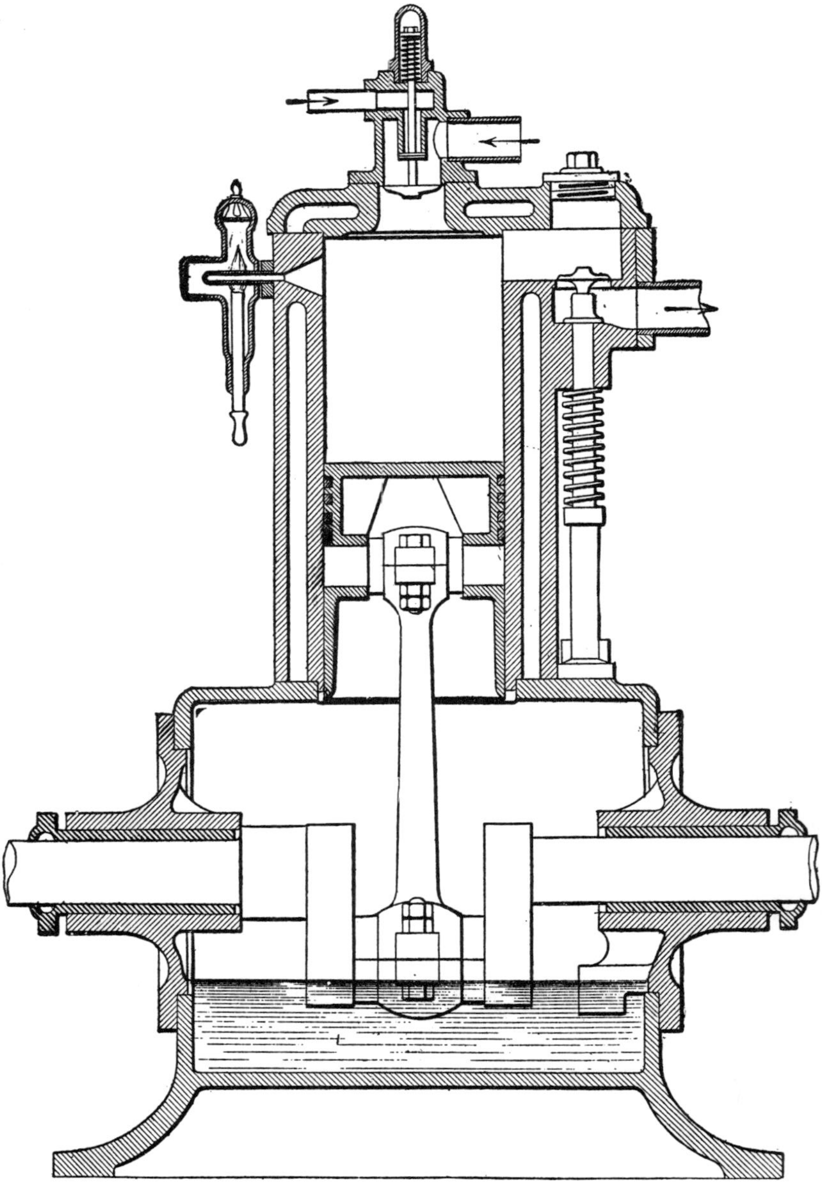

Fig. 37.—Section through an engine of the vertical or "steam-hammer" type.

can be directly coupled up with the crank-shaft of the engine, thus dispensing with a belt, which, at the least,

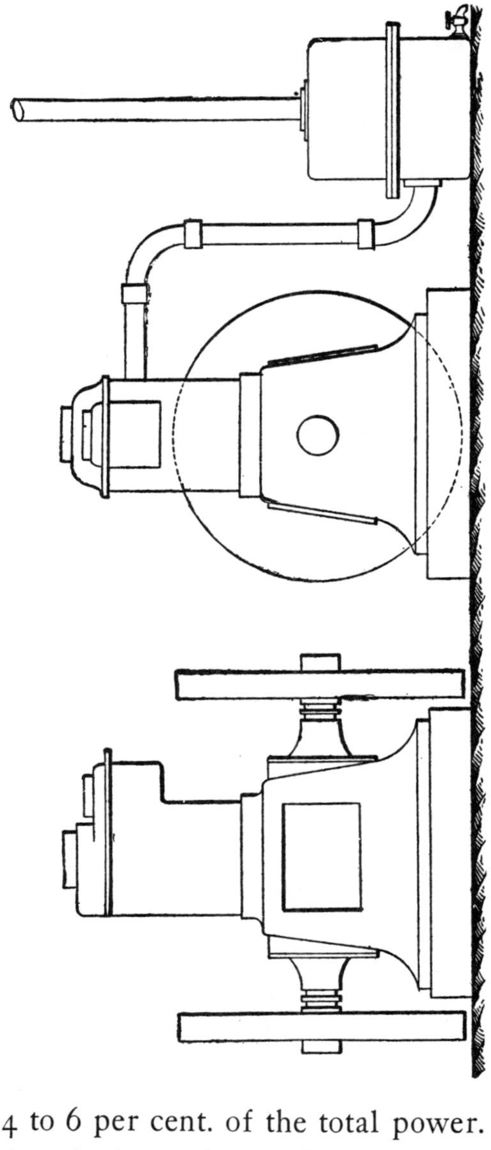

FIG. 38.—Side and end elevations of a vertical or "steam-hammer" engine.

absorbs 4 to 6 per cent. of the total power. It should, nevertheless, be borne in mind that the direct coupling of electric generators to engine-shafts implies the use of extremely large and, therefore, of extremely costly dynamos. Furthermore, by reason of this arrangement, groups of electro-generators can be disposed in a comparatively small amount of space. Some English manufacturers are also beginning to adopt the "steam-hammer" type of engine for high powers, the result being a marked saving in material and lowering of the cost of installation.

Power of the Engine.—The first thing to be considered is that the power of a gas-engine is always given in "effective" horse-power, and that the power of a steam-engine is always given in "indicated" horse-power in contracts of sale. In England and in the United States, the expression "nominal" horse-power is still employed. It may be advisable to define these various terms exactly, since unscrupulous dealers, to the buyer's loss, have done much to confuse them.

"Indicated" horse-power is a designation applied to the theoretical power produced by the action of the motive agent on the piston. The work performed is measured on an indicator card, by means of which the average pressure to be considered in the computation of the theoretical power is ascertained.

The "effective" or brake horse-power is equal to the "indicated" horse-power, less the energy absorbed by passive resistance, friction of the moving parts, etc.

The "effective" work is an experimental term applied to the power actually developed at the shaft. This work is of interest solely to the engine user.

In a well-built motor, in which the passive resistance by reason of the correct adjustment and simplicity of the parts, is reduced to a minimum, the "effective" horse-power is about 80 to 87 per cent. of the "indicated" horse-power, when the engine runs under full load. This reduced output is usually called the "mechanical efficiency" of the engine.

"Nominal" horse-power is an arbitrary term in the sense in which it is used in England and America, where it is quite common. The manufacturers themselves do not seem to agree on its absolute value. A "nominal" horse-power, however, is equal to anything from 3 to 4 "effective" horse-power. The uncertainty which ensues from the use of the term should lead to its abandonment.

In installing a motor, the determination of its horse-power is a matter of grave importance, which should not be considered as if the motor were a steam-engine or an engine of some other type. It must not be forgotten that, especially at full load, explosion-engines are most efficient, and that, under these conditions, it will generally be advisable to subordinate the utility of having a reserve power to the economy which follows from the employment of a motor running at a load close to its maximum capacity. On the other hand, the gas-engine user is unwilling to believe that the stipulated horse-power of the motor which is sold to him is the greatest that it is capable of developing under industrial conditions. Business competition has led some firms to sell their engines to meet these conditions. It is probably not stretching the truth too far to declare that 80 per cent. of the engines sold with no exact contract specifications are incapable of maintaining for more than a half hour the power which is attributed to them, and which the buyer expects. It follows that the power at which the engine is sold should be both industrially realized and maintained, if need be, for an entire day, without the engine's showing the slightest perturbation, or faltering in its silent and regular operation. To attain this end, it is essential that the energy developed by the engine in normal or constant operation should not exceed 90 to 95 per cent. of the maximum power which it is able to yield, and which may be termed its "utmost power." As a general rule, especially for installations in which the power fluctuates from the lowest possible to double this, as much attention must be paid to the consumption at half load as at full load; and preference should be given to the engine which, other things being equal, will operate most economically at its lowest load. In this case the consumption per effective horse-power is appreciably higher. Generally, this consumption is greater by 20 to 30 per cent. than that at full load. This is particularly true of the single-acting engines so widely used for horse-powers less than 100 to 150.

In some double or triple-acting engines, according to certain writers, the diminution in the consumption will hardly be proportional to the diminution of the power, or at any rate, the difference between the consumption per B. H. P. at full load and at reduced load will be less than in other engines. It should be observed, however, that this statement is apparently not borne out by experiments which the author has had occasion to make. To a slight degree, this economy is obtained at the cost of simplicity, and consequently, at the cost of the engine. At all events, the engines have the merit of great cyclic regularity, rendering them serviceable for driving electric-light dynamos; but this regularity can also be attained by the use of the extra heavy fly-wheels which English firms, in particular, have introduced.

Automatic Starting.—When the gas-engine was first introduced, starting was effected simply by manually turning the fly-wheel until steady running was assured. This procedure, altogether too crude in its way, is attended with some danger. In a few countries it is prohibited by laws regulating the employment of industrial machinery. If the engine be of rather large size—one, moreover, which operates at high pressure—such a method of starting is very troublesome. For these reasons, among others, manufacturers have devised automatic means of setting a gas-engine in motion.

Of such automatic devices, the first that shall be mentioned is a combination of pipes, provided with cocks, by the manipulation of which, a certain amount of gas, drawn from the supply pipe, is introduced into the engine-cylinder. The piston is first placed in a suitable position, and behind it a mixture is formed which is ignited by a naked flame situated near a convenient orifice. When the explosion takes place the ignition-orifice is automatically closed, and the piston is given its motive impulse. The engine thus started continues to run in accordance with the regular recurrence of the cycles. In this system, starting is effected by the explosion of a mixture, without previous compression.

Some designers have devised a system of hand-pumps which compress in the cylinder a mixture of air and gas, ignited at the proper time by allowing it to come into contact with the igniter, through the manipulation of cocks (Fig. 39).

These two methods are not absolutely effective. They require a certain deftness which can be acquired only after some practice. Furthermore, they are objectionable because the starting is effected too violently, and because the instantaneous explosion subjects the stationary piston, crank, and fly-wheel to a shock so sudden that they may be severely strained and may even break. Moreover, the slightest leakage in one of the valves or checks may cause the entire system to fail, and, particularly in the case of the pump, may induce a back explosion exceedingly dangerous to the man in charge of the engine.

These systems are now almost generally supplanted by the compressed-air system, which is simpler, less dangerous, and more certain in its effect.

The elements comprising the system in question include essentially a reservoir of thick sheet iron, capable of resisting a pressure of 180 to 225 pounds and sufficient in capacity to start an engine several times. This reservoir is connected with the engine by piping, which is disposed in one of two ways, depending upon whether

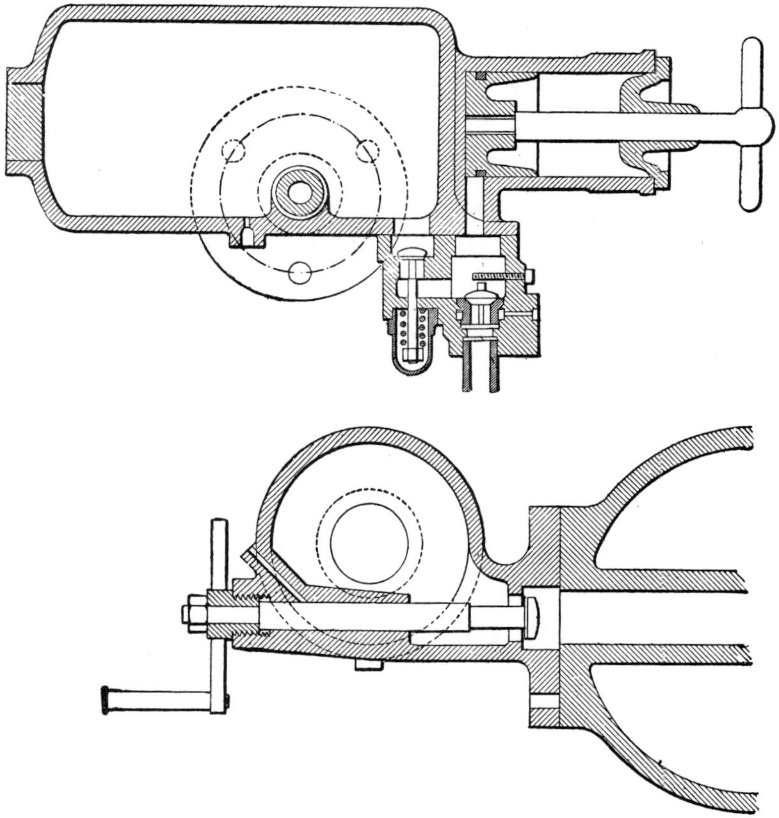

FIG. 39.—Tangye starter.

the reservoir is charged by the engine itself operatively connected with the compressor, or by an independent compressor, mechanically operated.

In the first case, the pipe is provided with a stop-cock, mounted adjacent to the cylinder, and with a check-valve. When the engine is started and the gas cut off, the air is drawn in at each cycle and driven back into the reservoir during the period of compression. When the engine, running under these conditions by reason of the inertia of the fly-wheel, begins to slow down, the check-valve is closed and the gas-admission valve opened, so as to produce several explosions and to impart a certain speed to the engine in order to continue the charging of the reservoir with compressed air. This done, the valve on the reservoir itself is tightly closed, as well as the check-valve, so as to avoid any leakage likely to cause a fall in the reservoir's pressure.

In the second case, which applies particularly to engines of more than 50 horse-power, the charging pipe connected with the reservoir is necessarily independent of the pipe by means of which the motor is started. The reservoir having been filled and the decompression cam thrown into gear, starting is accomplished:

1. By placing the piston in starting position, which corresponds with a crank inclination of 10 to 20 degrees in the direction of the piston's movement, from the rear dead center, immediately after the period of compression;

2. By opening the reservoir-valve;

3. By allowing the compressed air to enter the cylinder rapidly, through the quick manipulation of the stop-cock, which is closed again when the impulse is given and reopened at the corresponding period of the following cycle, this operation being repeated several times in order to impart sufficient speed to the motor;

4. By opening the gas-valve and finally closing the two valves of the compressed-air pipe.

The pipes and compressed-air reservoirs should be perfectly tight. The reservoirs should have a capacity in inverse ratio to the pressure under which they are placed, *i.e.,* they increase in size as the pressure decreases. If, for example, the reservoirs should be operated normally at a pressure of 105 to 120 pounds per square inch, their capacity should be at least five or six times the volume of the engine-cylinder. If these reservoirs are charged by the engine itself, the pressure will always be less by 15 to 20 per cent. than that of the compression.

CHAPTER III

THE INSTALLATION OF AN ENGINE

IN the preceding chapter the various structural details of an engine have been summarized and those arrangements indicated which, from a general standpoint, seem most commendable. No particular system has been described in order that this manual might be kept within proper limits. Moreover, the best-known writers, such as Hutton, Hiscox, Parsell and Weed, in America; Aimé Witz, in France; Dugald Clerk, Frederick Grover, and the late Bryan Donkin, in England; Güldner, Schottler, Thering, in Germany, have published very full descriptive works on the various types of engines.

We shall now consider the various methods which seem preferable in installing an engine. The directions to be given, the author believes, have not been hitherto published in any work, and are here formulated, after an experience of fifteen years, acquired in testing over 400 engines of all kinds, and in studying the methods of the leading gas-engine-building firms in the chief industrial centers of Europe and America.

Location.—The engine should be preferably located in a well-lighted place, accessible for inspection and maintenance, and should be kept entirely free from dust. As a general rule, the engine space should be enclosed. An engine should not be located in a cellar, on a damp floor, or in badly illuminated and ventilated places.

Gas-Pipes.—The pipes by which fuel is conducted to engines, driven by street-gas, and the gas-bags, etc., are rarely altogether free from leakage. For this reason, the engine-room should be as well ventilated as possible in the interest of safety. Long lines of pipe between the meter and the engine should be avoided, for the sake of economy, since the chances for leakage increase with the length of the pipe. It seldom happens that the leakage of a pipe 30 to 50 feet long, supplying a 30 horse-power engine, is much less than 90 cubic feet per hour. The beneficial effect of short supply pipes between meter and engine on the running of the engine is another point to be kept in mind.

An engine should be supplied with gas as cool as possible, which condition is seldom realized if long pipe lines be employed, extending through workshops, the temperature of which is usually higher than that of underground piping. On the other hand, pipes should not be exposed to the freezing temperature of winter, since the frost formed within the pipe, and particularly the crystalline deposition of naphthaline, reduces the cross section and sometimes clogs the passage. Often it happens that water condenses in the pipes; consequently, the piping should be disposed so as to obviate inclines, in which the water can collect in pockets. An accumulation of water is usually manifested by fluctuations in the flame of the burner. In places where water can collect, a drain-cock should be inserted. In places exposed to frost, a cock or a plug should be provided, so that a liquid can be introduced to dissolve the naphthaline. To insure the perfect operation of the engine, as well as to avoid fluctuations in nearby lights, pipes having a large diameter should preferably be employed. The cross-section should not be less than that of the discharge-pipe of the meter, selected in accordance with the prescriptions of the following table:

GAS-METERS.

Capacity.	Normal hourly flow.	Dimension in inches.			Diameter of pipe in inches.	Power of engine to be fed.
		Height.	Width.	Depth.		
3 burners	14.726 cu. ft.	13	11	$9\frac{13}{16}$	0.590	1/2 h.-p.
5 "	24.710 "	18	$13\frac{3}{4}$	$10\frac{5}{8}$	0.787	3/4 "
10 "	49.420 "	$21\frac{1}{4}$	$18\frac{1}{2}$	$12\frac{9}{16}$	0.984	1-2 "
20 "	98.840 "	$23\frac{3}{16}$	$19\frac{11}{16}$	$15\frac{5}{16}$	1.181	3-4 "
30 "	148.260 "	$25\frac{5}{8}$	$21\frac{11}{16}$	$18\frac{3}{16}$	1.456	5-6 "
50 "	247.100 "	$29\frac{1}{2}$	$24\frac{5}{16}$	$20\frac{7}{16}$	1.592	7-10 "
60 "	296.520 "	$30\frac{5}{16}$	$25\frac{5}{8}$	$25\frac{5}{8}$	1.671	11-14 "
80 "	395.360 "	$33\frac{5}{16}$	$30\frac{5}{16}$	$27\frac{1}{8}$	1.968	15-19 "
100 "	494.200 "	35	$33\frac{7}{16}$	$29\frac{15}{16}$	1.968	20-25 "
150 "	741.300 "	$40\frac{3}{16}$	$40\frac{3}{16}$	$33\frac{13}{16}$	—	30-40 "

The records made are exact only when the meters (Fig. 40) are installed and operated under normal conditions. Two chief causes tend to falsify the measurements in wet meters: (1) evaporation of the water, (2) the failure to have the meter level.

Evaporation occurs incessantly, owing to the flowing of the gas through the apparatus, and increases with a rise in the temperature of the atmosphere surrounding the meter. Consequently this temperature must be kept down, for which reason the meter should be placed as near the ground as possible. The evaporation also increases with the volume of gas delivered. Hence the meter should not supply more than the volume for which it was intended. In order to facilitate the return of the water of condensation to the meter and to prevent its accumulation, the pipes should be inclined as far as possible toward the meter. The lowering of

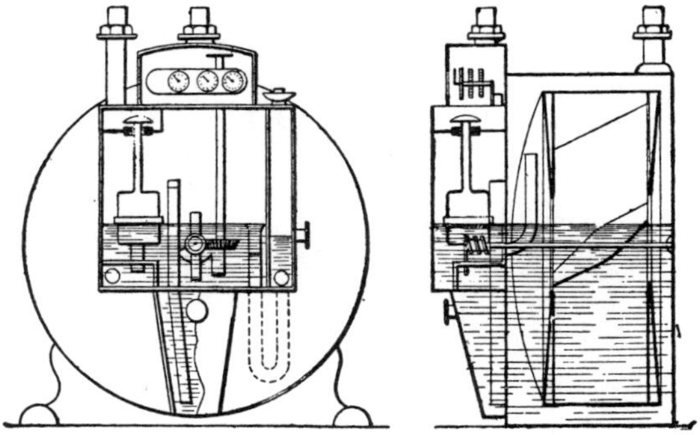

FIG. 40.—Wet gas-meter.

the water-level in the meter benefits the consumer at the expense of the gas company.

Inclination from the horizontal has an effect that varies with the direction of inclination. If the meter be inclined forward, or from left to right, the water can flow out by the lateral opening at the level, and incorrect measurements are made to the consumer's cost.

During winter, the meter should be protected from cold. The simplest way to accomplish this, is to wrap substances around the meter which are poor conductors of heat, such as straw, hay, rags, cotton, and the like. Freezing of the water can also be prevented by the addition of alcohol in the proportion of 2 pints per burner. The water is thus enabled to withstand a temperature of about 5 degrees F. below zero. Instead of alcohol, glycerine in the same proportions can be employed, care being taken that the glycerine is neutral, in order

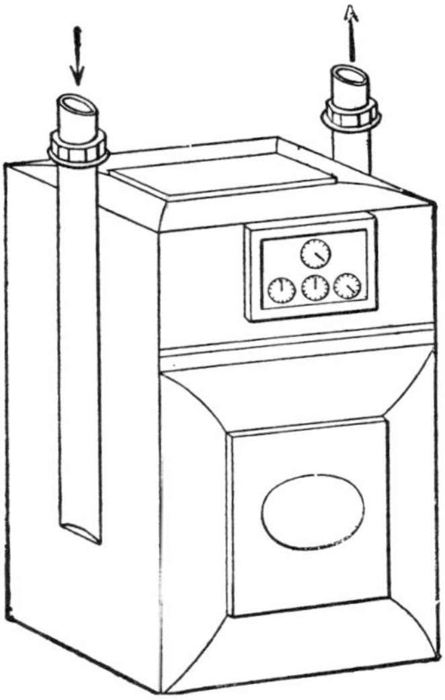

FIG. 41.—Dry gas-meter.

that the meter may not be attacked by the acids which the liquid sometimes contains.

Dry Meters.—Dry meters are employed chiefly in cold climates, where wet meters could be protected only with difficulty and where the water is likely to freeze. In the United States the dry meter is the type most widely employed. In Sweden and in Holland it is also generally introduced (Fig. 41).

In the matter of accuracy of measurement there is little, if any, difference between wet and dry meters. The dry meter has the merit of measuring correctly regardless of the fluctuations in the water level. On the other hand, it is open to the objection of absorbing somewhat more pressure than the wet meter, after having been in operation for a certain length of time.

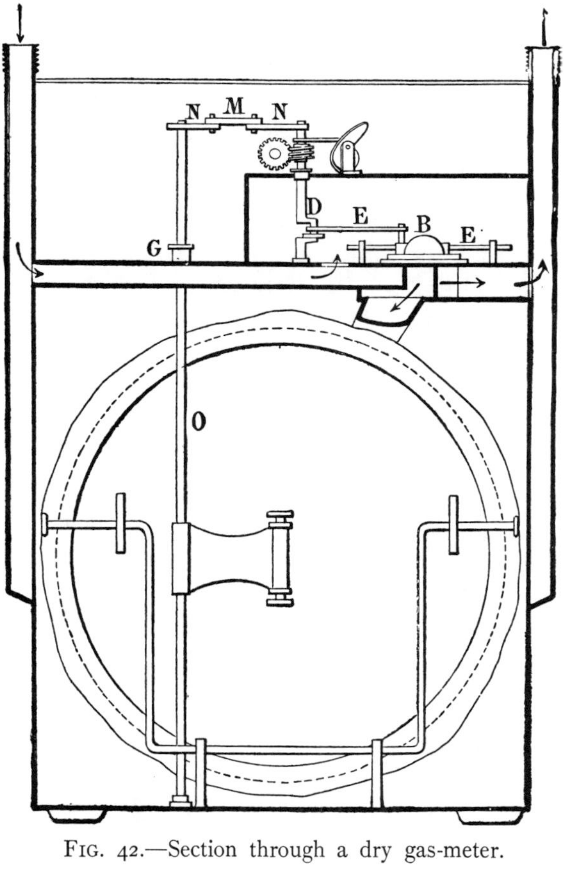

FIG. 42.—Section through a dry gas-meter.

This is an objection of no great weight; for there is always enough pressure in the mains and pipes to operate a meter.

In many cases, where the employment of non-freezing liquids is necessary, the dry meter may be used to advantage, since all such liquids have more or less corroding effect on sheet lead and even tin, depending upon the composition of the gas.

The dry meter comprises two bellows, operating in a casing divided into two compartments by a central

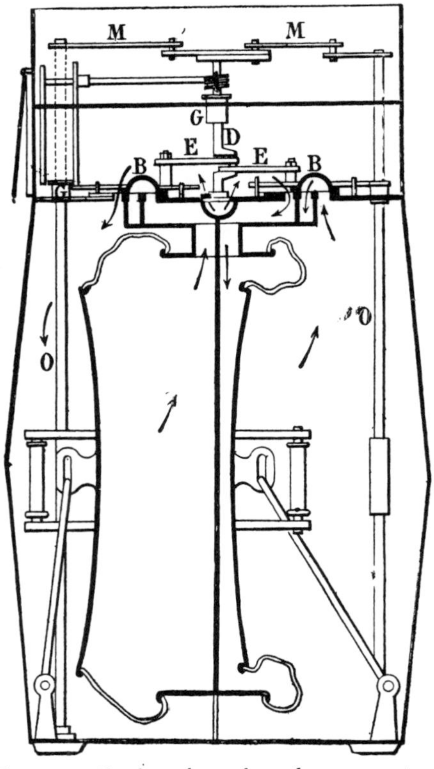

FIG. 43.—Section through a dry gas-meter.

partition. The gas is distributed on one or the other side of the bellows, by slides *B*. The slides *B* are provided with cranks *E*, controlled by levers *M*, actuated by transmission shafts *O*, driven by the bellows. The meter is adjusted by a screw which changes the throw of the cranks *E* and consequently affects the bellows. The movement of the crank-shaft *D* is transmitted to the indicating apparatus. In order to obviate any leakage, this shaft passes through a stuffing-box,

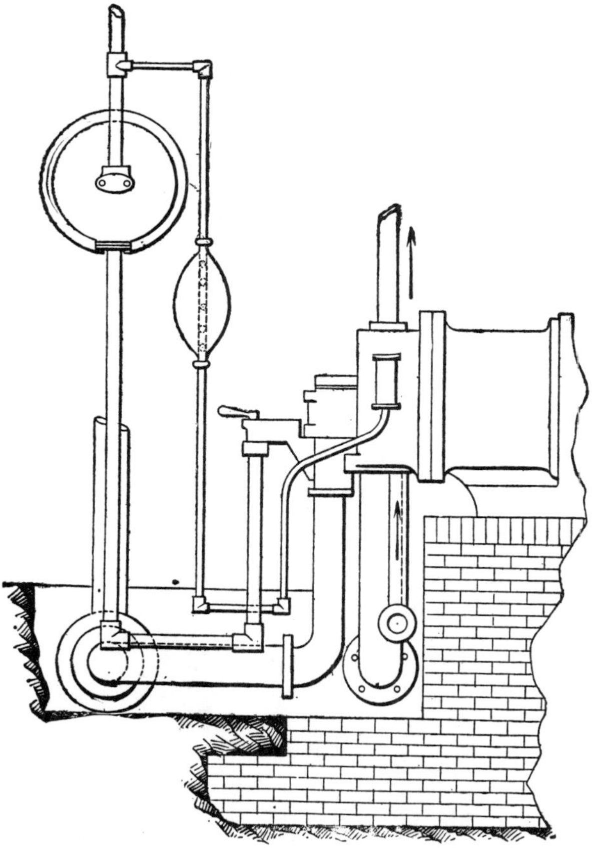

FIG. 44.—Rubber bag to prevent fluctuations of the ignition flame.

G. The diagrams (Figs. 42-43) show the construction of a dry meter, the arrows indicating the course taken by the gas.

Care should be taken to provide the gas-pipe with a drain-cock, at a point near the engine. By means of this cock, any air in the pipe can be allowed to

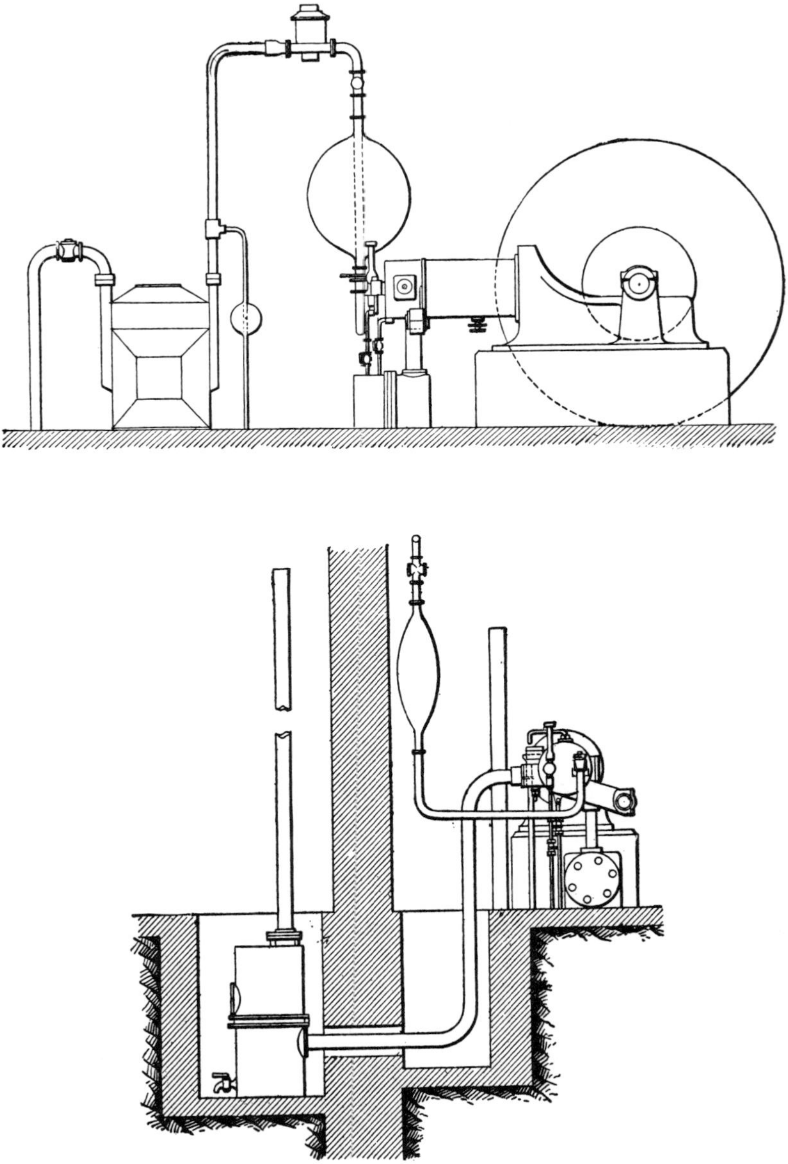

FIG. 45.—Rubber bags on gas-pipes.

escape before starting; otherwise the engine can be set in motion only with difficulty. If the engine be provided with an incandescent tube, the gas-supply pipe of the igniter should be fitted with a small rubber pouch or bag, in order to obviate fluctuations in the burner flame, caused by variations in the pressure (Fig. 44). As a general rule, the supply-pipe should be connected with the main pipe on the forward side of the bags and gas-governors. The main pipe and all other piping near the engine should extend underground, so that free access to the motor from all sides can be obtained, without possibility of injury.

Anti-pulsators, Bags, Pressure-Regulators.—The most commonly employed means of preventing fluctuation of nearby lights, due to the sharp strokes of the engine, consists in providing the gas-supply pipe with rubber bags (Fig. 45), which form reservoirs for the gas and, by reason of their elasticity, counteract the effect produced by the suction of the engine. Nevertheless, in order to insure a supply of gas at a constant pressure, which is necessary for the perfect operation of the engine, there are generally used, in addition to the bags, devices called gas-governors, or anti-pulsators (Fig. 46).

Although these devices are constructed in different ways, the underlying principle is the same in all. They comprise a metallic casing, containing a flexible diaphragm of rubber or of some fabric impermeable to gas. Suction of the engine creates a vacuum in the casing. The diaphragm bends, thereby actuating a valve, which cuts off the gas supply. During the three following periods (compression, explosion, and exhaust) the gas, by reason of its pressure on the diaphragm, opens the valve and fills the casing, ready for the next suction stroke.

Other devices, which are never sold with the engine, but are rendered necessary by reason of the conditions imposed by the gas supply are sold under the name

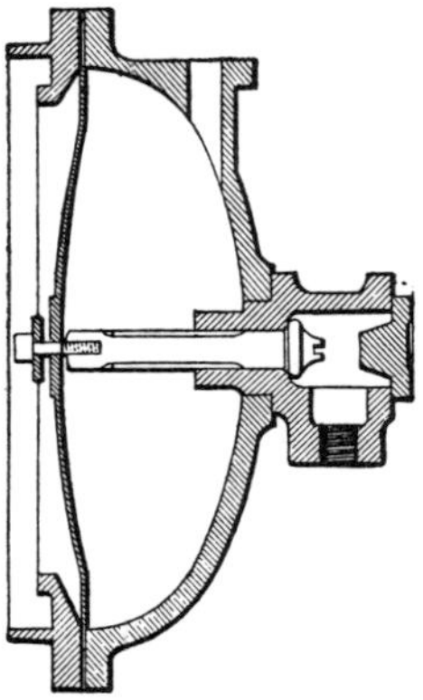

FIG. 46.—An anti-pulsator.

"pressure-regulators" (Fig. 47). They consist of a bell, floating in a reservoir containing water and glycerine (or mercury), and likewise actuate a valve which partially controls the flow of gas. This valve being balanced, its mechanical action is the more certain. Such devices are very effective in maintaining the steadiness of lights. On the other hand, they are often an obstacle to the operation of the engine because they reduce the flow and pressure of the gas too much. In order to obviate this difficulty, a pressure-regulator should be chosen with discrimination, and of sufficiently large size to insure the maintenance of an adequate supply of gas to the engine. Frequent examinations should be made to ascertain if the bell of the regulator is immersed in the liquid. In the case of anti-pulsators, care should be taken that they are not spattered with oil, which has a disastrous effect on rubber. Anti-pulsators are generally mounted about 4

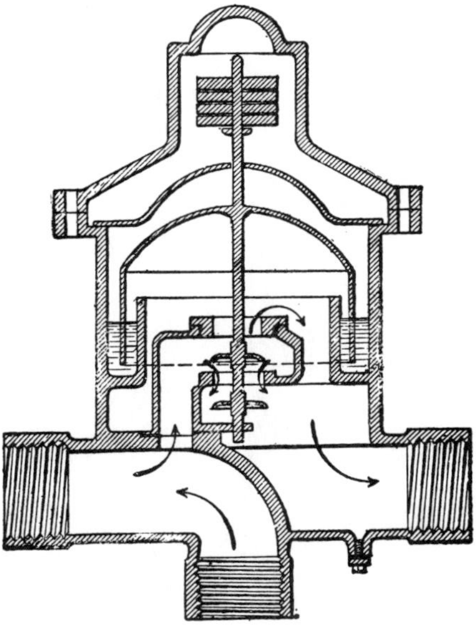

FIG. 47.—A pressure-regulator.

inches from a wall, in order that the diaphragm may be actuated by hand, if need be.

Precautions.—In order not to strain the rubber of the bags or of the anti-pulsators, it is advisable to place a stop-cock in advance of these devices so that they can not be filled while the motor is at rest.

The capacity of the rubber bags that can be bought

in the market being limited, it is necessary to place one, two, or three extra bags in series (Figs. 48 and 49), for large pipes; but it should be borne in mind that the

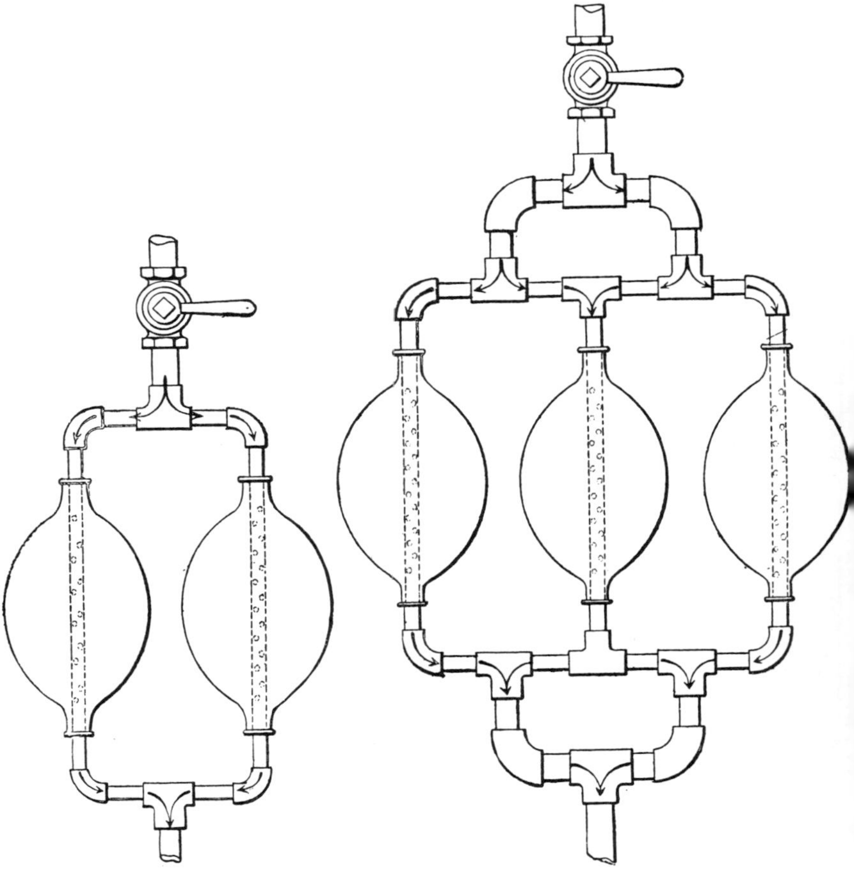

FIGS. 48-49.—Arrangement of rubber bags.

total section of the branch pipes should be at least equal to that of the main pipe. It is also advisable to extend the tube completely through the bag as shown in Figs. 48 and 49.

If there be two branch pipes the minimum diameter

which meets this requirement is ascertained as follows: Draw to any scale a semicircle having a diameter equal or proportional to that of the main pipe (Fig. 50). The sides of the isosceles triangle inscribed within this semicircle give the minimum diameter of each of the branch pipes.

Sometimes engines are provided with a cock having an arrangement by means of which the gas feed is permanently regulated, according to the quality and pres-

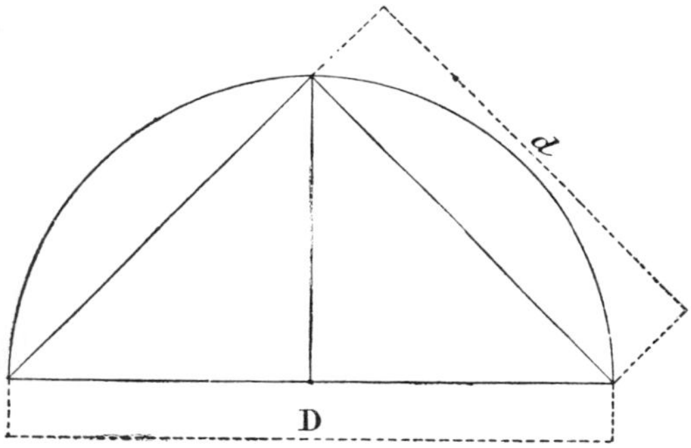

FIG. 50.

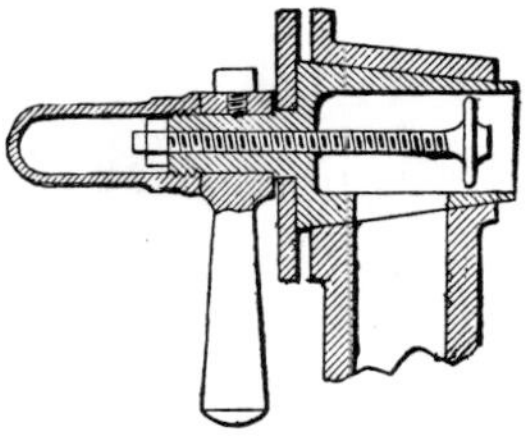

FIG. 51.

sure of the gas and according to the load at which the engine is to run. This renders it possible to open the cock always to the same point (Fig. 51).

Air Suction.—In a special chapter the precautions to be taken to counteract the influence of the suction of the engine in causing vibration will be treated. The manner in which the suction of air is effected necessarily has as marked an influence on the operation of the engine as the supply of gas, since air and gas constitute the explosive mixture.

Resistance to the suction of air should be carefully

avoided, for which reason the length of the pipe should be reduced to a minimum, and its cross-section kept at least equal to that of the air inlet of the engine. Since the quality of street-gas varies with each city, the proper proportions of gas and air are not constant. In order that these proportions may be regulated, it is a matter of some importance to fit some suitable device on the pipe. Good engines are provided with a plug or flap valve. Generally the air-pipe terminates either in the hollowed portion of the frame, or in an independent pot, or air chest. The first arrangement is not to be recommended for engines over 20 to 25 horse-power. Accidents may result, such as the breaking of the frame by reason of back firing, of which more will be said later. If an independent chest be employed, its closeness to the ground renders it possible for dust easily to pass through the air-holes in the walls at the moment of suction, and even to enter the cylinder, where its presence is particularly harmful, leading, as it does, to the rapid wear of the rubbing surfaces. This evil can be largely remedied by filling the air-chest with cocoa fiber or even wood fiber, provided the latter does not become packed down so as to prevent the air from passing freely. Such fibers act as air-filters. Regular cleaning or renewal of the fiber protects the cylinder from wear. In a general way, care should be taken, before fitting both the gas and air pipes, to tap the pipes, elbows, and joints lightly with a hammer on the outside in order to loosen whatever rust or sand may cling to the interior; otherwise this foreign matter may enter the cylinder and cause perturbations in the operation of the engine. Under all circumstances, care should be taken not to place the end of the air-pipe under the floor or in an enclosed space, because leakage may occur, due to the bad seating of the air-valve, thereby producing a mixture which may explode if the flame leaps back, as we shall see in the discussion of suction by pipes terminating in the hollow of the frame. On the other hand, sand or saw-dust should not be sprinkled on the floor.

Exhaust.—For the exhaust, cast-iron or drawn pipes as short as possible should be used. Not only the power of the engine, but also its economic consumption, can be markedly affected by the employment of long and bent pipes. Resistance to the exhaust of the products of combustion not only causes an injurious counter-pressure, but also prevents the clearing of the cylinder of burnt gases, which contaminate the aspired mixture and rob it of much of its explosiveness. The necessity of evacuating the cylinder as completely as possible is, nevertheless, not always reconcilable with local surroundings. To a certain extent, the objections to long exhaust-pipes are overcome by rigorously avoiding the use of elbows. Gradual curves are preferable. In the case of very long pipes it is advisable to increase their diameter every 16 feet from the exhaust. The exhaust-chest should be placed as near as possible to the engine; it should never be buried; for the joints of the inlet and outlet pipes of the exhaust-chest should be easily accessible, so that they may be renewed when necessary. The author recommends the placing of the exhaust-chest in a masonry pit, which can be closed with a sheet-metal cover. For engines of 20 horse-power and upward, these joints should be entirely of asbestos. Pipes screwed directly into the casting are liable to rust. Ex-

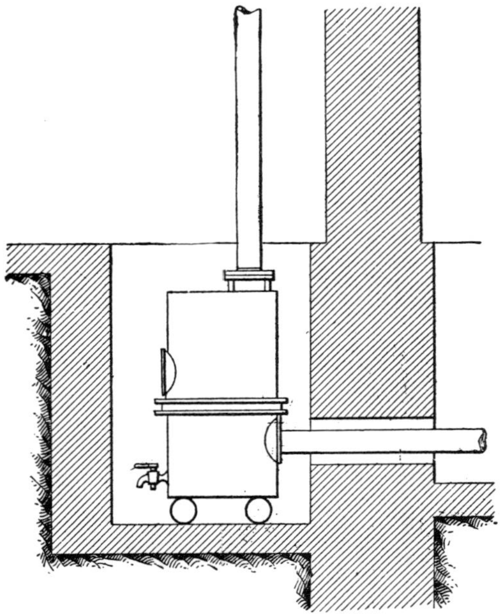

FIG. 52.—Method of mounting pipes.

posed as they are to the steam or water of the exhaust, they cannot be detached.

The water, which results from the combination of the hydrogen of the gas with the oxygen of the air, is deposited in most cases at the bottom of the exhaust-chest. It is advisable to fit a plug or iron cock in the base of the chest. Alkaline or acid water will always corrode a bronze cock. In order that the pipes may not also be attacked, they are not disposed horizontally, but are given a slight incline toward the point where the water is drained off. If pipes of some length be employed, they should be able to expand freely without straining the joints, as shown in the accompanying diagram (Fig. 52), in which the exhaust-chest rests on iron rollers which permit a slight displacement.

For the sake of safety, at least that portion of the piping which is near the engine should be located at a proper distance from woodwork and other combustible material. By no means should the exhaust discharge into a sewer or chimney, even though the sewer or chimney be not in use; for the unburnt gases may be trapped, and dangerous explosions may ensue at the moment of discharge.

The joints or threaded sleeves employed in assembling the exhaust-pipe should be tested for tightness. The combined action of the moisture and heat causes the metal to rust and to deteriorate very rapidly at leaky spots.

When several engines are installed near one another, each should be provided with a special exhaust-pipe; otherwise it may happen, when the engines are all running at once, that the products of combustion discharged by the one may cause a back pressure detrimental to the exhaust of the next.

It is possible to employ a pipe common to all the exhausts if the pipe starts from a point beyond the exhaust-chests, in which case Y-joints and not T-joints are to be used.

The manner of securing the pipes to walls by means of detachable hangers, lined with asbestos, is shown in a general way in the accompanying Fig. 53. The object of this arrangement is to render detachment easy and to prevent the transmission of shocks to the masonry.

The precautions to be taken for muffling the noise of the exhaust will be discussed later.

The end of the exhaust-pipe should be slightly

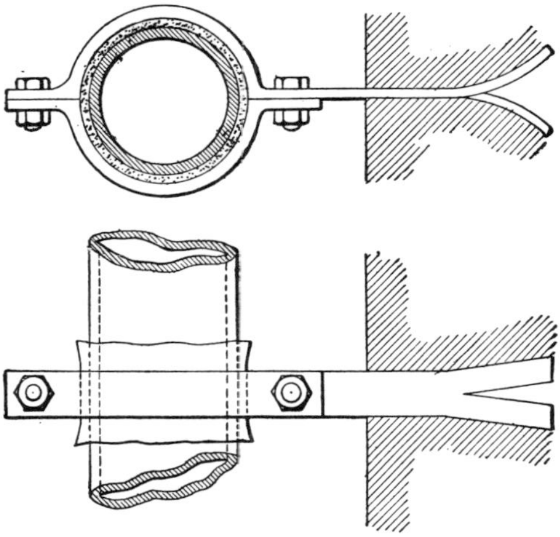

FIG. 53.—Method of securing pipes to walls.

curved down in order to prevent the entrance of rain. Exhaust-pipes are subjected to considerable vibration, due to the sudden discharge of the gases. To protect the joints, the pipes should be rigidly fastened in place.

Legal Authorization.—In most countries gas-engines may be installed only in accordance with the provision of general or local laws, which impose certain conditions. These laws vary with different localities, for which reason they are not discussed here.

CHAPTER IV

FOUNDATION AND EXHAUST

THE reader will remember from what has already been said that a gas-engine is a motor which, more than any other, is subjected to forces, suddenly and repeatedly exerted, producing violent reactions on the foundation. It follows that the foundation must be made particularly resistant by properly determining its shape and size and by carefully selecting the material of which it is to be built.

The Foundation Materials.—Well-hardened brick should be used. The top course of bricks should be laid on edge. It is advisable to increase the stability of the foundation by longitudinally elongating it toward the base, as shown in the accompanying diagram (Fig. 54).

As a binding material, only mortar composed of coarse sand or river sand and of good cement, should be used. Instead of coarse sand, crushed slag, well-screened, may be employed. The mortar should consist of ⅔ slag and ⅓ cement. Oil should not in any way come into contact with the mortar; it may percolate through the cement and alter its resistant qualities.

As in the construction of all foundations, care should be taken to excavate down to good soil and to line the bottom with concrete, in order to form a single mass of artificial stone. A day or two should be allowed for the masonry to dry out, before filling in around it.

When the engine is installed on the ground floor above a vaulted cellar, the foundation should not rest

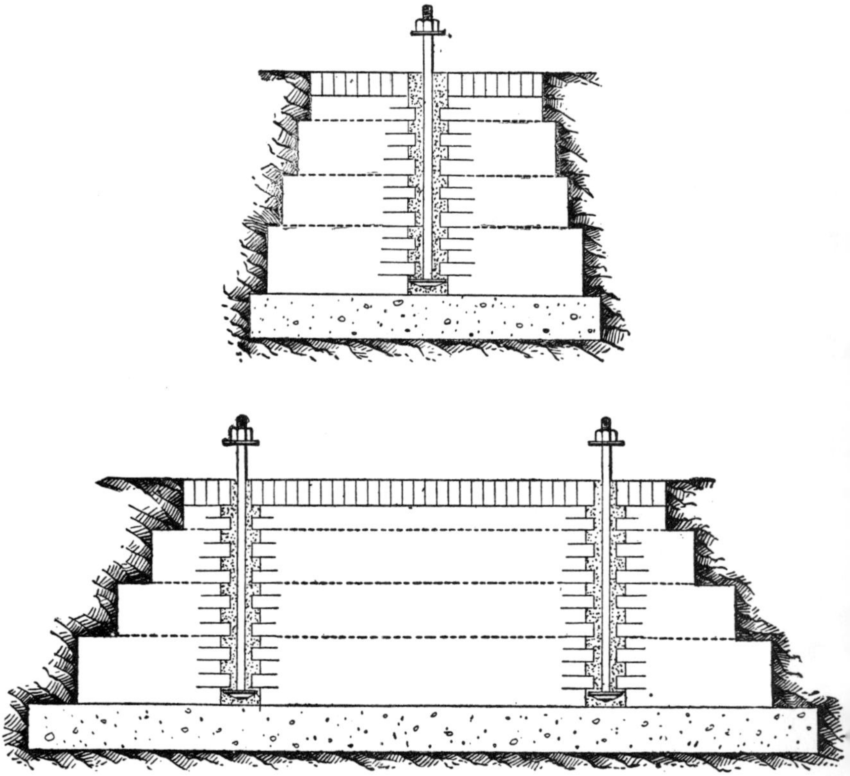

FIG. 54.—Method of building the foundation.

directly on the vault below or on the joists, but should be built upon the very floor of the cellar, so that it passes through the planking of the ground floor without contact.

When the engine is to be installed on a staging, the

method of securing it in place illustrated in Fig. 55 should be adopted.

Although a foundation, built in the manner described, will fulfill the usual conditions of an industrial installation, it will be inadequate for special cases in which trepidation is to be expected. Such is the case when engines are to be installed in places where, owing to the absence of factories, it is necessary to avoid

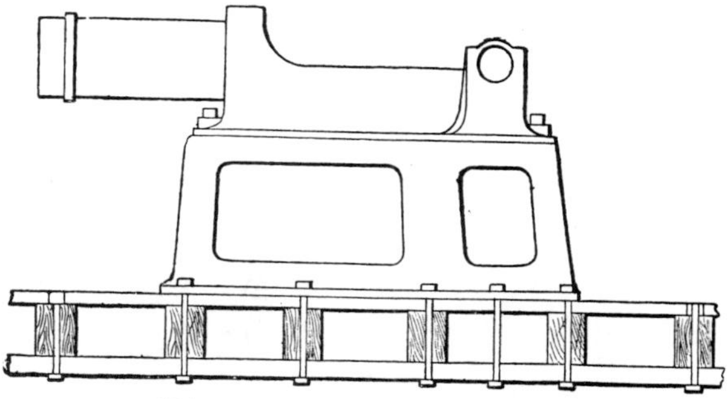

FIG. 55.—Elevated foundation.

all nuisance, such as noise, trepidations, odors, and the like.

Vibration.—In order to prevent the transmission of vibration, the foundation should be carefully insulated from all neighboring walls. For this purpose various insulating substances called "anti-vibratory" are to be recommended. Among these may be mentioned horsehair, felt packing, cork, and the like. The efficacy of these substances depends much on the manner in which they are applied. It is always advisable to interpose a layer of one of these substances, from one to four inches thick, between the foundation and the surrounding soil, the thickness varying with the nature of the material used and the effect to be obtained. Between the bed of concrete, mentioned previously, and the foundation-masonry and between the foundation and the engine-frame, a layer of insulating material may well be placed. Preference is to be given to substances not likely to rot or at least not likely to lose their insulating property, when acted upon by heat, moisture or pressure.

Here it may not be amiss to warn against the utilization of cork for the bottom of the foundation; for water may cause the cork to swell and to dislocate the foundation or destroy its level.

The employment of the various substances mentioned does not entail any great expense when the foundations are not large and the engines are light. But the cost becomes considerable when insulating material is to be employed for the foundation of a 30 to 50 horse-power engine and upwards. For an engine of such size the author recommends an arrangement as simple as it is efficient, which consists in placing the foundation of the engine in a veritable masonry basin, the bottom of which is a bed of concrete of suitable thickness. The foundation is so placed that the lateral surfaces are absolutely independent of the supporting-walls of the basin thus formed. Care should be taken to cover the bottom with a layer of dry sand, rammed down well, varying in thickness with each case. This layer of sand constitutes the anti-vibratory material and confines the trepidations of the engine to the foundation.

As a result of this arrangement, it should be observed that, being unsupported laterally, the foundation should be all the more resistant, for which reason the base-area and weight should be increased by 30 to 40 per cent. The expense entailed will be largely offset by saving the cost of special anti-vibratory substances. In places liable to be flooded by water, the basin should be cemented or asphalted.

When the engine is of some size and is intended for the driving of one or more dynamos which may themselves give rise to vibrations, the dynamos are secured directly to the foundation of the engine, which is extended for that purpose, so that both machines are carried solidly on a single base.

The foregoing outline should not lead the proprietor of a plant to dispense with the services of experts, whose long experience has brought home to them the difficulties to be overcome in special cases.

It should here be stated, as a general rule, that the bricks should be thoroughly moistened before they are laid in order that they may grip the mortar.

After having been placed on the foundation and roughly trimmed with respect to the transmission devices, the engine is carefully leveled by means of hardwood wedges driven under the base. This done, the bolts are sealed by very gradually pouring a cement wash into the holes, and allowing it to set. When the holes are completely filled and the bolts securely fastened in place, a shallow rim, or edge of clay, or sand is run around the cast base, so as to form a small box or trough, in which cement is also poured for the purpose of firmly binding the engine frame and foundation together. When, as in the case of electric-light engines, single extra-heavy fly-wheels are employed, provided with bearings held in independent cast supports, the following rule should be observed to prevent the overheating due to unlevelness, which usually occurs at the bushings of these bearings: That part of the foundation which is to receive such a support should rest directly on the concrete bed and should be rigidly connected at the bottom with the main foundation. When the foundation is completely blocked up, the fly-wheel bearing with its support is hung to the crank-shaft; and not until this is effected is the masonry at the base of the support completed and rigidly fixed in its proper position.

For very large engines, the foundation-bolts should be particularly well sealed into the foundation. In order to attain this end the bricks are laid around the bolt-holes, alternately projected and retracted as shown in Fig. 54. Broken stone is then rammed down around the fixed bolt; in the interstices cement wash is poured.

Air Vibration, etc.—Vibration due chiefly to the transmission of noises and the displacement of air by the piston should not be confused with the trepidations previously mentioned.

The noise of an engine is caused by two distinct phenomena. The one is due to the transmitting properties of the entire solid mass constituting the frame, the foundation, and the soil. The other is due to vibrations transmitted to the air. In both cases, in order to reduce the noise to a minimum, the moving parts should be kept nicely adjusted, and above all, shocks avoided, the more harmful of which are caused by the play between the joint at the foot of the connecting-rod and the piston-pin, and between the head of the connecting-rod and the crank-shaft.

Although smooth running of the engine may be assured, there is always an inherent drawback in the rapid reciprocating movement of the piston. In large, single-acting gas-engines, a considerable displacement of air is thus produced. In the case of a forty horse-power engine having a cylinder diameter and piston-stroke respectively of $13\frac{3}{4}$ inches and $21\frac{3}{5}$ inches, it is evident that at each stroke the piston will displace about 2 cubic feet of air, the effect of which will be doubled when it is considered that on the forward stroke back pressure is created and on the return stroke suction is produced.

The air motion caused by the engine is the more readily felt as the engine-room is smaller. If the room, for example, be 9 feet by 75 feet by 8 feet, the volume will be 1,080 cubic feet. From this it follows that the 2 cubic feet of air in the case supposed will be alternately displaced six times each second, which means the displacement of 12 cubic feet at short intervals with an average speed of 550 feet per minute. Such vibrations transmitted to halls or neighboring rooms are due entirely to the displacement of the air.

In installations where the air-intake of the engine is

located in the engine-room, a certain compensation is secured, at the period of suction, between the quantity of air expelled on the forward stroke of the piston and the quantity of air drawn into the cylinder. From this it follows that the vibration caused by the movement of the air is felt less and occurs but once for two revolutions of the engine.

This phenomenon is very manifest in narrow rooms in which the engine happens to be installed near glass windows. By reason of the elasticity of the glass, the windows acquire a vibratory movement corresponding in period with half the number of revolutions of the engine. It follows from the preceding that, in order to do away with the air vibration occasioned by the piston in drawing in and forcing out air in an enclosed space, openings should be provided for the entrance of large quantities of air, or a sufficient supply of air should be forced in by means of a fan.

The author ends this section with the advice that all pipes in general and the exhaust-pipe in particular be insulated from the foundation and from the walls through which they pass as well as from the ground, as metal pipes are good conductors of sound and liable to carry to some distance from the engine the sounds of the moving parts.

Exhaust Noises.—Among the most difficult noises to muffle is that of the exhaust. Indeed, it is the exhaust above all that betrays the gas-engine by its discharge to the exterior through the exhaust-pipe. The most commonly employed means for rendering the exhaust less perceptible consists in extending the pipe upward as far as possible, even to the height of the roof. This is an easy way out of the difficulty; but it has a bad effect on the operation of the engine. It reduces the power generated and increases the consumption, as will be explained in a special paragraph.

Expansion-boxes, more commonly called exhaust-mufflers, considerably deaden the noise of explosion by

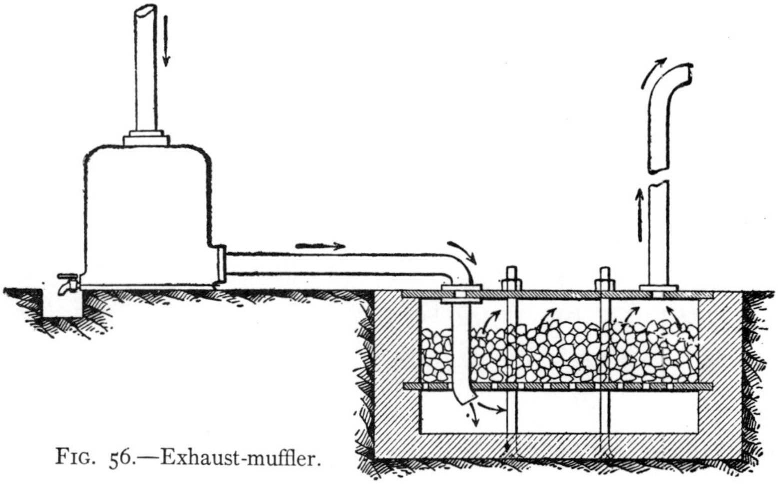

FIG. 56.—Exhaust-muffler.

the use of two or three successive receptacles. But this remedy is attended with the same faults that mark the use of extremely long pipes. The best plan is to mount a single exhaust-muffler near the discharge of the engine in the engine-room itself, where it will serve at least the purpose of localizing the sound.

The employment of pipes of sufficiently large cross-section to constitute expansion-boxes in themselves will also muffle the exhaust. A more complete solution of the problem is obtained by causing the exhaust-pipe, after leaving the muffler, to discharge into a masonry trough having a volume equal to twelve times that of the engine-cylinder (Fig. 56). This trough should be divided into two parts, separated by a horizontal iron grating. Into the lower part, which is empty, the exhaust-pipe discharges; in the upper part, paving-blocks or hard stones not likely to crumble with the heat, are placed. Between this layer of stones and the cover it is advisable to leave a space equal to the first. Here the gases may expand after having been divided into many parts in passing through the spaces left between adjacent stones. The trough should not be closed by a rigid cover; for, although efficient muffling may be attained, certain disadvantages are nevertheless encountered. It may happen that in a badly regulated engine, unburnt gases may be discharged into this trough, forming an explosive mixture which will be ignited by the next explosion, causing considerable damage. Still, the explosion will be less dangerous than noisy. It may be mentioned in passing that this disadvantage occurs rarely.

A second arrangement consists in superposing the end of the exhaust-pipe upon a casing of suitable size, which casing is partitioned off by several perforated baffle-plates. This casing is preferably made of wood, lined with metal, so that it will not be resonant. The size of the casing, the number of partitions and their perforations, and the manner of disposing the partitions have much to do with the result to be obtained. Here again the experience of the expert is of use.

Various other systems are employed, depending upon the particular circumstances of each case. Among these systems may be mentioned those in which the pipe is forked at its end to form either a yoke (Fig. 57) or a double curve, each branch of which terminates in a muffler (Fig. 58).

It should be observed that, under ordinary conditions, noises heard as hissing sounds are often due to the

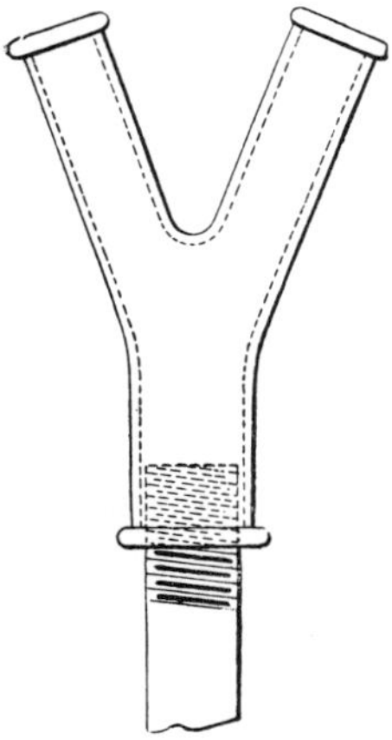

FIG. 57.

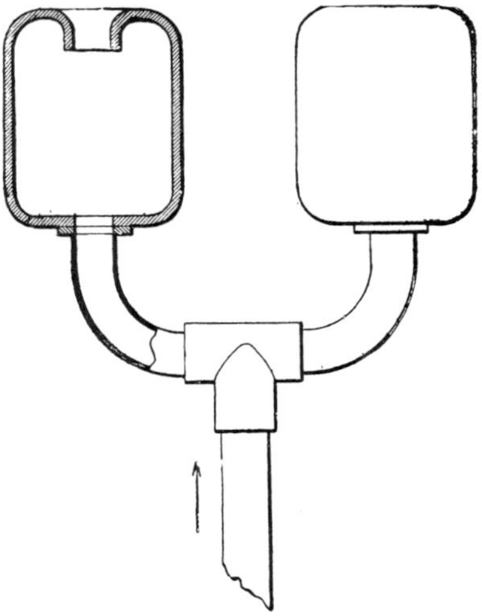

FIG. 58.

Two types of exhaust-mufflers.

presence of projections, or to distortion of the pipes near the discharge opening. Consequently, in connecting the pipes, care should be taken that the joints or seams have no interior projections. Occasionally, water may be injected into the exhaust-muffler in order to condense the vapors of the exhaust, the result being a deadening of the noises; but in order to be truly efficient this method should be employed with discretion, for which reason the advice of an expert is of value.

CHAPTER V

WATER CIRCULATION

CIRCULATION of water in explosion-engines is one of the essentials of their perfect operation. Two special cases are encountered. In the one the jacket of the engine is supplied with running water; in the other, reservoirs are employed, the circulation being effected simply by the difference in specific gravity in a thermosiphon apparatus. Coolers are also used.

Running Water.—A water-jacket fed from a constant source of running water, such as the water mains of a town, is certainly productive of the best results, the supply, moreover, being easily regulated; but the system is not widely used because the water runs away and is entirely lost. If running water be employed, the outlet of the jacket is so disposed that the water gushes out immediately on leaving the cylinder, and that the flow is visible and accessible, in order that the temperature may be tested by the hand. Apart from the relatively great cost of water in towns, the use of running water is objectionable on account of its chemical composition. Though it may be clear and limpid, it frequently contains lime salts, carbonates, sulphates, and silicates which are precipitated by reason of the sudden change of temperature to which the water is subjected as it comes into contact with the walls of the cylinder. That part of the water-jacket surrounding the head or explosion-chamber, where the temperature is necessarily the highest, becomes literally covered with calcareous incrustations, which are the more harmful because they are bad conductors of heat and because they reduce and even obstruct the passage exactly at the point where the water must circulate most freely to do any good. If the circulating water be pumped into the jacket, it is preferable, wherever possible, to use cistern water, which is not likely to contain lime salts in suspension. If river water be used, it should be free from the objections already mentioned, which are all the more grave if the water be muddy, as sometimes happens. The water-jacket can be easily freed from all non-adhering deposits by flushing it periodically through the medium of a conveniently placed cock. It is always preferable to pass the water through a reservoir where its impurities can settle, before it flows to the cylinder. In the case considered, the water usually has an average temperature of 54 to 60 degrees F., under which condition the hourly flow should be at least 5½ gallons per horse-power per hour, the temperature rising at the outlet-pipe of the cylinder to 140 and 158 degrees F., which should not be surpassed. However, in engines working with high compression, 104 to 122 degrees F. should not be exceeded.

If the water-jacket be fed by a reservoir, it is essential that the reservoir comply with the following conditions:

In horizontal engines the water-inlet is always located in the base of the cylinder, while the outlet is located at the top. By providing the inlet-pipe extending to the cylinder with a cock, the circulation of water can be regulated to correspond with the work performed by the engine. Another cock at the end of the outlet-pipe near the reservoir serves, in conjunction

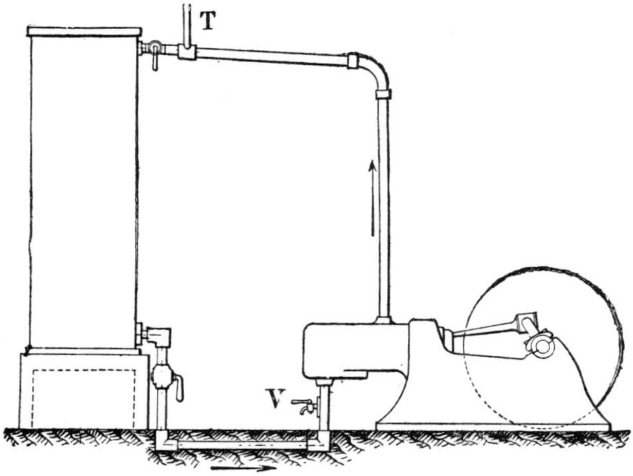

FIG. 59.—Thermo-syphon cooling system.

with the first, to arrest the circulating water. When the weather is very cold or when the cylinder must be repaired, these two cocks may be closed, and the pipe and water-jacket of the cylinder drained by means of the drain-cock *V* (Fig. 59), mounted at the inlet of the engine's water-jacket. In order that the pressure of the atmosphere may not prevent the flowing of the water, the highest part of the pipe is provided with a small tube, *T*, communicating with the atmosphere.

On account of the importance of preventing losses

of the charge in the pipes the author recommends the utilization of sluice-valves of the type shown in Fig. 60, instead of the usual cone or plug type.

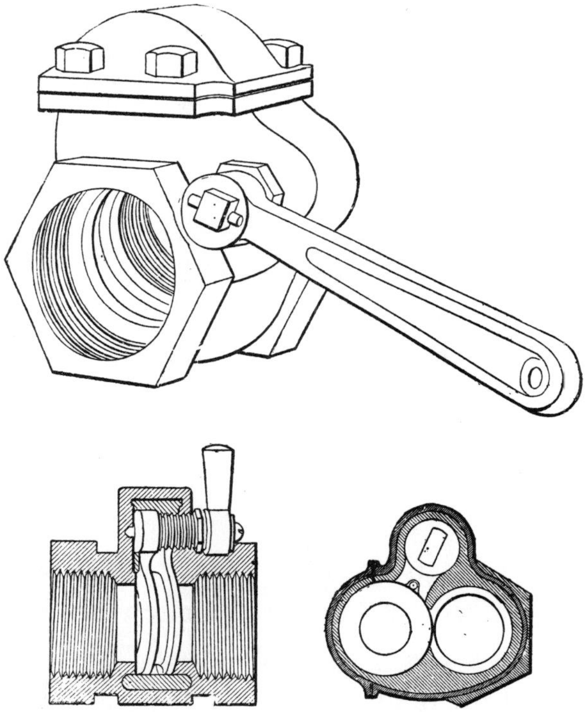

FIG. 60.—Vanne sluice-cock.

Water-Tanks.—The reservoir is mounted in such a way that its base is flush with the top of the cylinder; it should be as near as possible to the cylinder in order to obviate the use of long inlet and return pipes. This fact, however, does not necessarily render it advisable to place the reservoir in the engine-room; for such a disposition is doubly disadvantageous in so far as it does not permit a sufficiently rapid cooling of the circulating water by reason of the high temperature of the surrounding air, and in so far as it is liable to cause the formation of vapors which injuriously affect the engine. Consequently, the reservoir should be placed in as cool a place as possible, preferably even in the open air; for the water is not likely to freeze, except when it has been allowed to stand for a considerable time. The reservoir should be left uncovered so as to facilitate cooling by the liberation of the vapors formed on the surface of the water.

Circulation being effected solely by the difference in specific gravity or density between the warmer water emerging from the cylinder and the cooler water which flows in from the reservoir, the slightest obstruction will impede the flow. Hence, the cross-section of the pipes should not be less than that of the inlet and outlet openings of the cylinder of the engine. Good circulation cannot be attained if the water must overcome inclines or obstacles in the pipes themselves. Instead of elbows, long curves of great radius, limited to the smallest possible number, should be employed. This is particularly true of the return-pipe extending from the cylinder back to the reservoir. For this pipe a minimum incline of 10 to 15 per cent. should be allowed, in order that the water may run up into the reservoir. The height of the water in the reservoir should be from 2 to 4 inches above the discharge of the return-pipe. In order to maintain this level it is advisable to use some automatic device such as a float-valve, in which case the reservoir should not be allowed to become too full.

The size of a reservoir is determined by the engine; it should be large enough to enable the engine to run

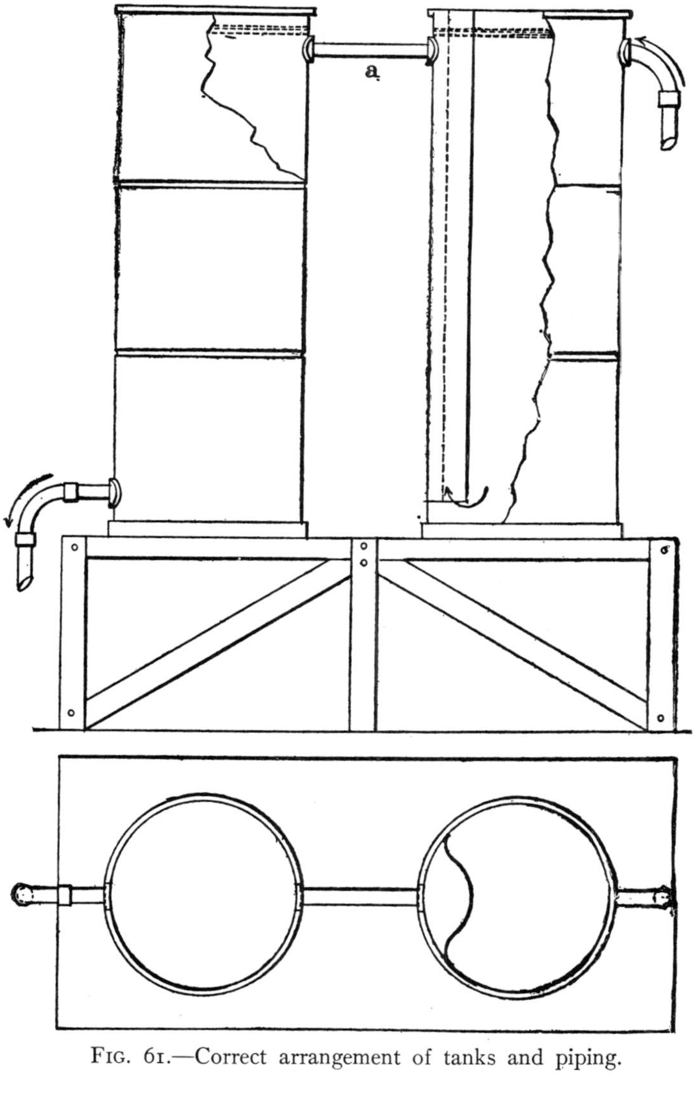

FIG. 61.—Correct arrangement of tanks and piping.

smoothly at its maximum load for several hours consecutively. Under these conditions, the reservoir should have a capacity of 45 to 55 gallons per horse-power for engines with "hit-and-miss" admission, and 55 to 65 gallons for engines controlled by variable admission. It is not advisable to employ reservoirs hav-

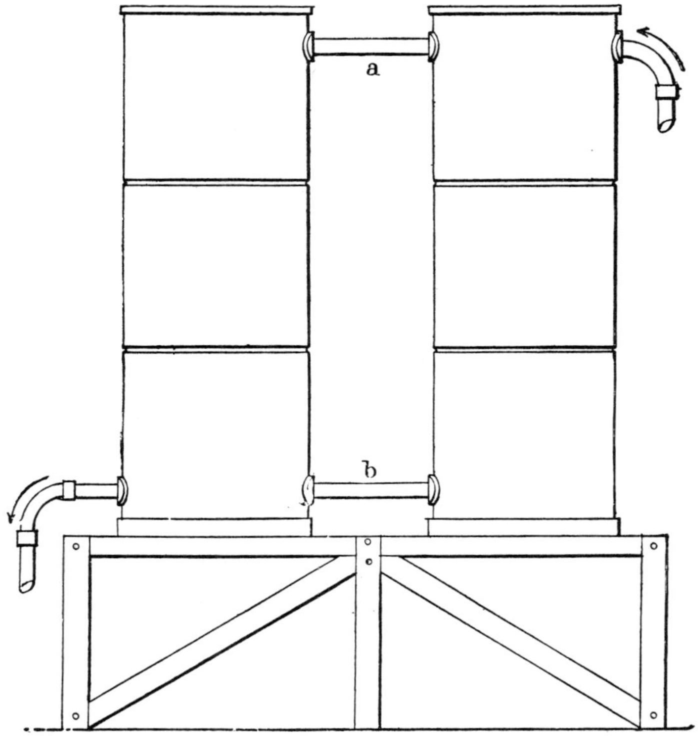

FIG. 62.—Incorrect arrangement of tanks and piping.

ing a capacity of more than 330 to 440 gallons, the usual diameter being about 3 feet.

If the power of the engine be such that several reservoirs are necessary, then the reservoirs should be connected in such a manner that the top of the first communicates with the bottom of the next and so on, the first reservoir receiving the water as it comes from the cylinder (Fig. 61).

Intercommunication of the reservoirs by means of a common top tube *(a)* is objectionable; and simultaneous intercommunication at top and bottom *(a* and *b)* is ineffective, so far as one of the reservoirs is concerned (Fig. 62).

The reservoirs are true thermo-siphons. Consequently the water should be methodically circulated; in other words, the hottest water, flowing from the en-

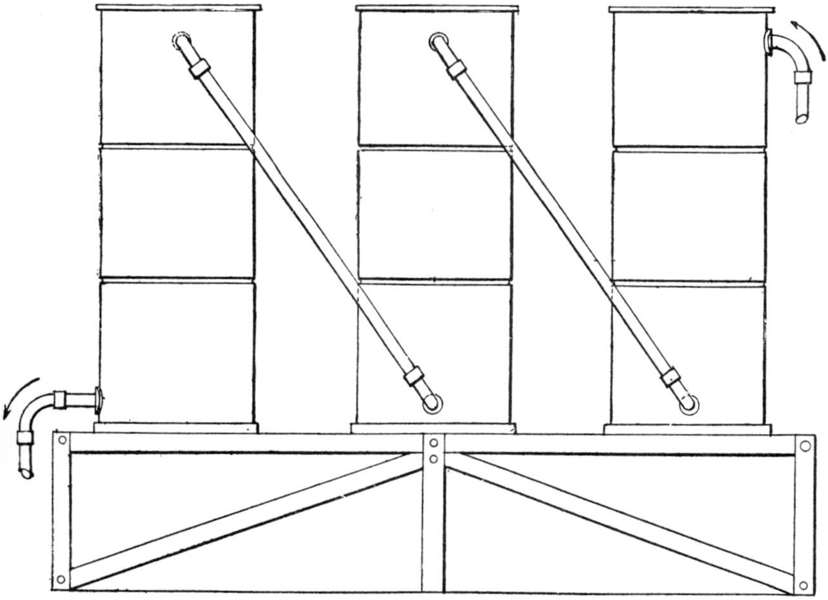

FIG. 63.—Tanks connected by inclined pipes.

gine into the top of the first reservoir and having, for example, a temperature of 104 degrees F., is cooled off to 86 degrees F. and drops to the bottom of the reservoir, thence to be driven, at a temperature sensibly equal to 86 degrees F., to the second reservoir, where a further cooling of 18 degrees F. takes place. In passing on to the following reservoirs the temperature is still further lowered, until the water finally reaches its minimum temperature, after which it flows back to the engine-cylinder.

In order to effect this cooling, the reservoirs can be connected in several ways. The most common method, as shown in Fig. 63, consists in connecting the reservoirs by oblique pipes. This is open to criticism, how-

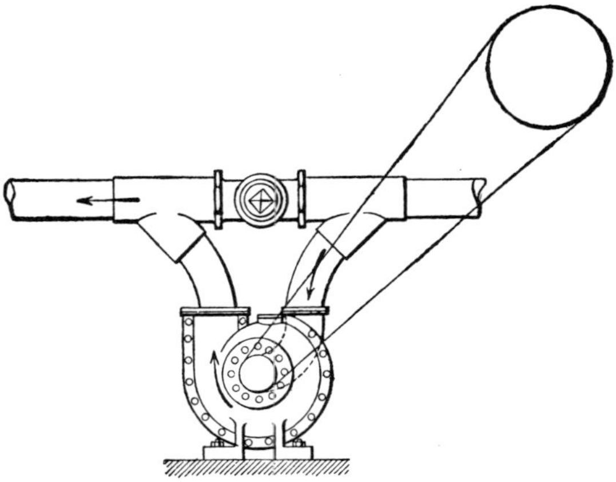

FIG. 64.—Circulating pump with by-pass.

ever, since leakage occurs, caused by the employment of elbows which retard the circulation. A less cumbrous and more efficient method of connection consists in joining the reservoirs by a single pipe at the top, as shown in Fig. 61; but care must be taken to extend this pipe at the point of its entrance into the adjoining reservoir by means of a downwardly projecting extension, or to fit its discharge-end with a box, closed by a single partition, open at the bottom.

In order to prevent incrustation of the water-jacket surrounding the cylinder, a pound of soda per 17 cubic feet of the reservoir capacity is monthly introduced, and the jacket flushed weekly by a cock conveniently mounted near the cylinder (Fig. 59). The jacket is thus purged of calcareous sediments, which are prevented by the soda from adhering to the metal. The flushing-cock mentioned also serves to drain the water-jacket of the cylinder in case of intense or persistent cold, which would certainly freeze the water in the jacket, thereby cracking the cylinder or the exposed pipes.

In order to regulate the circulation of the water in accordance with the work performed by the engine, a cock should be fitted to the water supply pipe at a convenient place.

In engines of large size, driven at full load for long periods, cooling by natural circulation is often inadequate. In such cases, circulation is quickened by a small rotary or reciprocating pump, driven from the engine itself and fitted with a by-pass provided with a cock. This arrangement permits the renewal of the natural thermo-siphon circulation in case of accident to the pump (Fig. 64).

Coolers.—The arrangement which is illustrated in Fig. 65, and which has the merit of simplicity, will be found of service in cooling the water. It comprises a tank *B* surmounted by a set of trays *E*, formed of frames to which iron rods are secured, spaced 1 to 2 feet apart, so as to form superimposed series separated by 1½ to

2 1/3 feet. On these trays bundles of tree branches are placed. The cold water at the bottom of the tank is forced by the pump *P* into the water-jacket, from which it emerges hot, and flows through the pipe *T*,

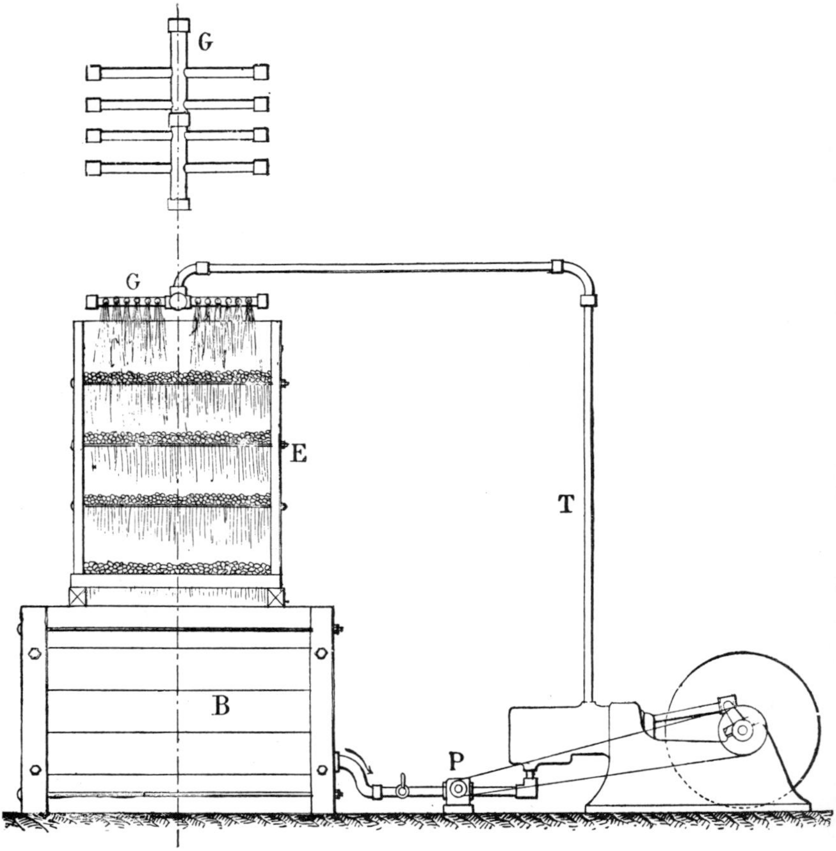

FIG. 65.—Water-cooler in which tree branches are employed.

which ends in a sprinkler *G*, formed of communicating tubes and perforated with a sufficient number of holes to enable the water to fall upon the trays in many drops. Thus finely divided, the water falls from one tray to another, retarded as it descends by the bundles of tree branches. It finally reaches the tank in a very cold condition and is then ready to be pumped to the engine. Birch branches are to be preferred on account of their tenuity.

Great care should be taken to cover the tank with a sheet-metal closure in order to prevent twigs and

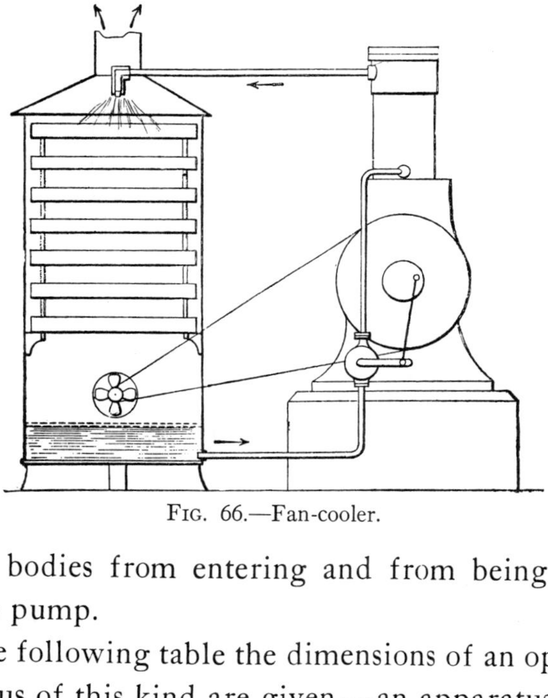

FIG. 66.—Fan-cooler.

foreign bodies from entering and from being drawn into the pump.

In the following table the dimensions of an operative apparatus of this kind are given,—an apparatus, moreover, that may be constructed of wood or of iron:—

Horse-power.	Volume in cubic ft.	Tank.		Height of tray-base.	Pump—Capacity in gals. per min.
		Base.	Height.		
30	105	4.9′ x 4.9′	4.4′	6.6′	16.71
40	154	5.2′ x 5.2′	5.6′	7.4′	18.69
50	190	5.7′ x 5.7′	6.4′	8.1′	21.99
75	350	6.6′ x 6.6′	8.1′	9.1′	35.18
100	490	7.4′ x 7.4′	9.1′	9.1′	43.98

In order that the water may not drop to one side, the base of the apparatus should be made 10 to 12 inches less in width than the tank.

The size of these apparatus may be considerably reduced by constructing them in the form of closed chests, into the bottom of which air may be injected by means of fans in order to accelerate cooling (Fig. 66).

CHAPTER VI

LUBRICATION

LUBRICATION is a subject that should be studied by every gas-engine user. So far as the piston is concerned it is a matter of the utmost importance. The piston does its work under very peculiar conditions. It is driven at great linear velocities; and it is, moreover, subjected to high temperatures which have nothing in common with good lubrication if care be not exercised.

The piston is the essential, vital element of an engine. Upon its freedom from leakage depends the maintenance of a proper compression, and, consequently, the production of power and economical consumption. As it travels forward and as it recedes from the explosion-chamber, it uncovers more and more of the frictional surface constituting the interior wall of the cylinder. This surface, as a result, is regularly brought into contact with the ignited, expanding gases after each explosion. For this reason the oil which covers the wall is constantly subjected to high temperatures, by which it is likely to be volatilized and burned. Therefore, the first condition to be fulfilled in properly lubricating the piston is a constant and regular supply of oil.

Quality of Oils.—For cylinder lubrication only the very best oils should be used; perfect lubrication is of such importance that cost should not be considered. Besides, the surplus oil which is usually caught in the drip-pan is by no means lost. After having been filtered it can be used for lubricating the bearings of the crank, the cam-shaft, and like parts.

Cylinder-oil should be exceedingly pure, free from acids, and composed of hydrocarbons that leave no residue after combustion. Only mineral oils, therefore, are suitable for the purpose. Those oils should be selected which, with a maximum of viscosity, are capable of withstanding great heat without volatilizing or burning. The point at which a good cylinder-oil ignites should not be lower than 535 degrees F.

Whether an oil possesses this essential quality is easily enough ascertained in practice without resorting to laboratory tests. All that is necessary is to heat the oil in a metal vessel or a porcelain dish. In order that the temperature may be uniform the vessel is shielded from the direct flame by interposing a piece of sheet metal or a layer of dry sand. As soon as gases begin to arise a lighted match is held over the oil. When the gases are ignited the thermometer reading is taken, the instrument being immersed in the oil. The temperature recorded is that corresponding with the point of ignition.

For cylinder lubrication American mineral oil is preferable to Russian oil. The specific gravity should lie somewhere between .886 and .889 at 70 degrees F.

Oil of this quality begins to evaporate at about 365 degrees F. Ignition occurs at 535 degrees F. The point of complete combustibility lies between 625 and 645 degrees F. Oil of this quality solidifies at 39 or 41 degrees F. Its color is a reddish yellow with a greenish fluorescence. Compared with water its degree of viscosity lies between 11.5 and 12.5 at a temperature of 140 degrees F.

Before lubricating other parts of the engine with oil that has been used for the piston, heavy particles and foreign matter, such as dust, bearing incrustations, and the like, should be filtered out. The piston-pivot and the connecting-rod head are preferably lubricated with fresh oil, because their constant movement renders inspection difficult and the control of lubrication irksome. A good, industrial mineral oil of usual market quality will be found satisfactory.

In order to bring home the importance of employing good cylinder-oil and of proper lubrication the author can only state that in his personal experience he has frequently detected losses varying from 10 to 15 per cent. in the power developed by engines poorly lubricated.

Types of Lubricators.—Among the more common apparatus employed for automatically lubricating the cylinder, the author mentions an English oiler of the type pictured in Fig. 67 which is driven simply by a belt from the intermediary shaft, and which rotates the pulley *P* secured on the shaft *a* of the apparatus, at a very slow speed. The shaft *a* is provided at its end with a small crank, from which a small iron arm *f* is suspended, which arm dips in the oil contained in the cup *G* of the oiler. When the shaft *a* is turned this arm, as it sweeps through the oil-bath, collects a certain quantity of oil which it deposits on the collector *b*. From this spindle the oil passes through an outlet-pipe opening into the bottom of the oiler, and thence to the cylinder. The entire apparatus is closed

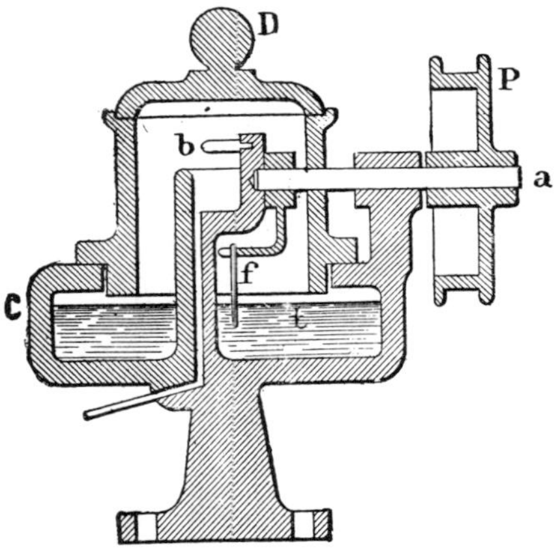

FIG. 67.—An automatic English oiler.

by a cover *D* which can be easily removed in order to ascertain the quantity of oil still remaining in the apparatus. Many other systems are utilized which, like the one that has been described, enable the feed to be controlled. Often small force-pumps are employed as cylinder-lubricators. Whatever may be the type selected, preference should be given to that in which the feed is visible (Fig. 68).

If the oil be fed under pressure the cylinder is more constantly lubricated. Pressure-lubricators are nowadays widely used on large engines. It is advisable to add a little salt to the water contained in sight-feed lubricators so that the drop of oil is easily freed.

These oil-pumps are provided with small check-valves at their outlets as well as at the inlets of cylinders. In order that pressure-lubricators may operate perfectly they should be regularly inspected and the check-valves ground from time to time.

The lubrication of the crank-shaft and of the two connecting-rod heads should receive every attention.

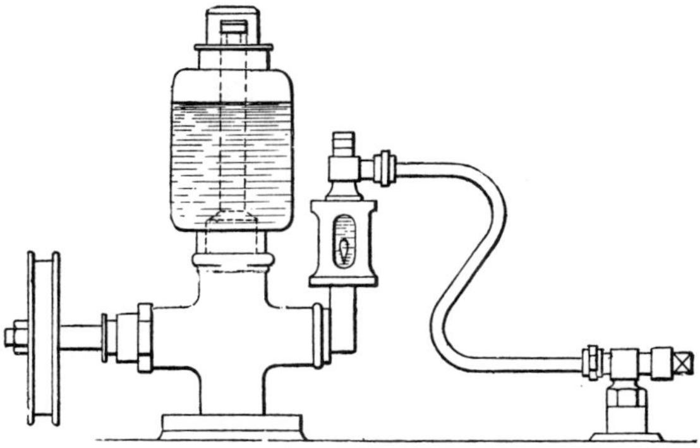

FIG. 68.—Sight-feed lubricating-pump.

Lubricating devices should be employed which, besides being efficient, do not necessitate the stopping of the engine in order to oil the bearings. The foot of the connecting-rod at the point where it is pivoted to the piston is generally lubricated with cylinder-oil which is supplied by a tube mounted in the proper place across the piston-wall (Fig. 69). This arrangement may be adequate enough for small engines; but it is not sufficiently sure for engines of considerable size. An independent lubricating system should be employed, lubrication being effected either by a splasher mounted in front of the cylinder or by a lubricator secured to the connecting-rod by which the pivot is lubricated through the medium of a small tube supplying special oil (Fig. 21). The head of the connecting-rod where it meets the crank, must also be carefully lubricated because of the important nature of the work which it must perform, and because of the shocks to which it is subjected at each explosion. For

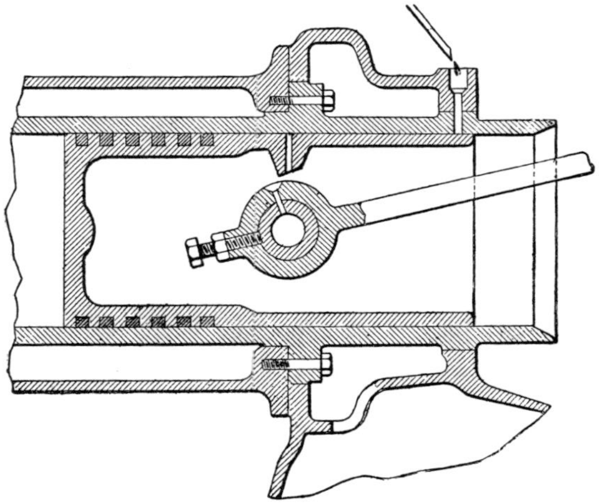

FIG. 69.—Method of oiling the piston and end of the connecting-rod.

motors of high power the system which seems to give most satisfactory results is that illustrated in Fig. 70. The arrangement there shown consists of an annular vessel secured at one side of the crank and turning concentrically on its axis; the vessel being connected with a long tube extending into a channel formed in the crank and discharging at the surface of the crank-pin within the bearing at the head of the connecting-rod. An adjustable sight-feed lubricator conducts the oil along a pipe to the vessel. Turning with the shaft, the vessel retains the oil in the periphery so that the feed in the previously mentioned channel in the connecting-rod head, is constant.

The main crank-shaft bearings are more easily lubricated. Among the systems commonly used with good results may be mentioned that shown in Fig. 71, in which the half section represents a small tube starting

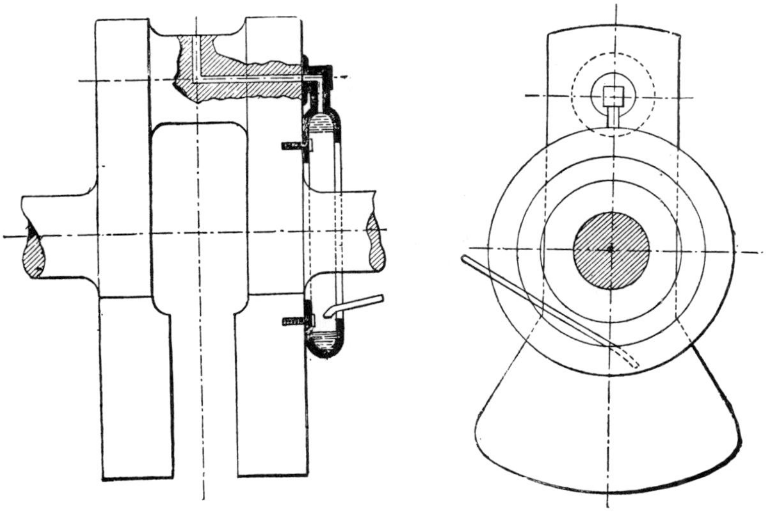

FIG. 70.—Method of oiling the crank-shaft.

from the bearing and terminating in the interior of an oil recess or reservoir cast integrally with the bearing-cap. This reservoir is filled up to the level of the tube opening. A piece of cotton waste held on a small iron wire is inserted in the tube, part of the cotton being allowed to hang down in the reservoir. This cotton serves as a kind of siphon and feeds the bearing by capillary attraction with a constant quantity of oil, the supply being regulated by varying the thickness of the cotton. When the motor is stopped, the cotton should be removed in order that oil-feeding may not uselessly continue. Glass, sight-feed lubricators with stop-cocks, are very often used on crank-shafts. They are cleaner and much more easily regulated. Of all shaft-bearing lubricators, those which are most to be recommended are of the revolving-ring type (Fig. 72). They presuppose, however, bearings of large

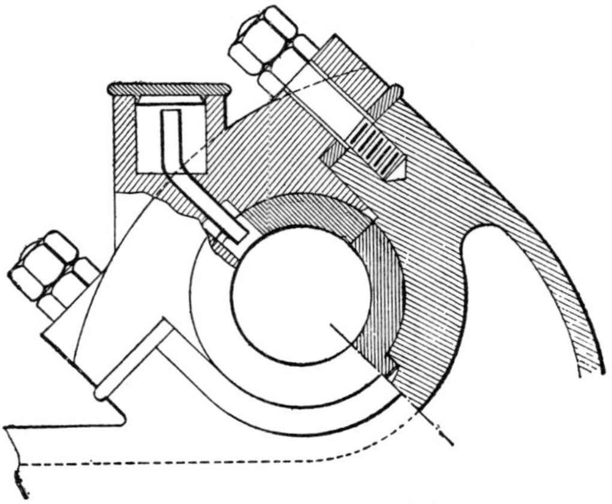

FIG. 71.—Cotton-waste lubricator.

size and a special arrangement of bushings which renders their application somewhat expensive. Furthermore, the revolving-ring system can hardly be used in connection with engines of less than 20 horse-power. Since the system is applied almost exclusively to dynamo-shafts, it need not here be described in detail. As its name indicates, it consists of a metal ring having a diameter larger than that part of the shaft from which it is suspended and by which it is rotated. The lower part of the ring is immersed in an oil bath so that a certain quantity of lubricant is continually transferred to the shaft.

The revolving ring bearing should be fitted with a drain-cock and a glass tube in order to control the level of the oil in the bearing.

Many manufacturers have adopted lubricating de-

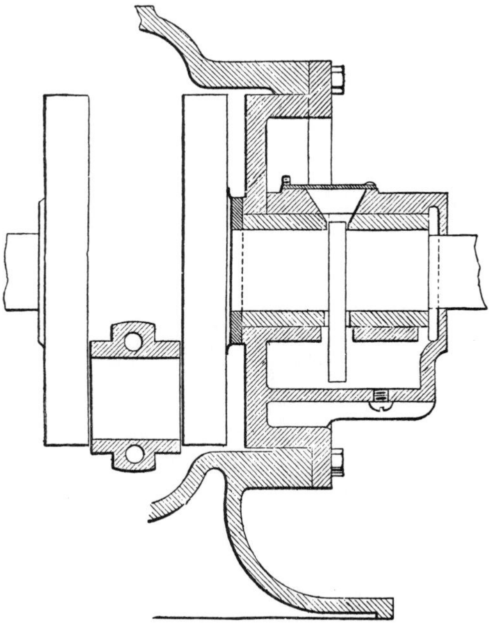

FIG. 72.—Ring type of bearing oiler.

vices for valve-stems, and especially for exhaust-valves. The system adopted consists of a small tube curved in any convenient direction and discharging in the stem-guide. The free end is provided with a plug. A few drops of petroleum are introduced once or twice a day.

The lubrication of an engine entails certain difficulties which are easily overcome. One of these is the splashing of oil by the connecting-rod head. In order that this splashed oil may be collected in the base of the engine a suitably curved sheet-metal guard is mounted over the crank. A more serious difficulty is presented when the oil from a crank-bearing finds its way to the hub of the fly-wheel, whence it is driven by the centrifugal force to the rim. The oil is not only splashed against the walls of the engine-room, but it also destroys the adhesion of the belt if the fly-wheel

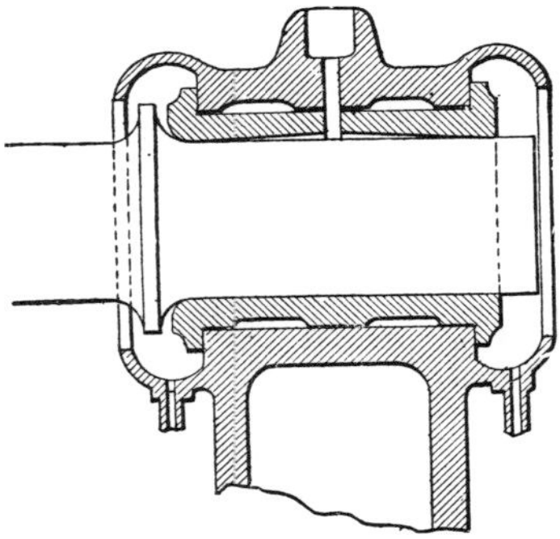

FIG. 73.—Shaft with oil-guard.

be employed as a pulley. In order to overcome this objection the oil is prevented from spreading along the shaft by means of a circular guard (Fig. 73) mounted on that portion of the shaft toward the interior of the bearing.

The problem of lubrication is of particular importance if the engine is driven for several days at a time without a stop. This happens in the case of mill and shop engines. Lubricators of large volume or lubricators which can be readily filled without stopping the engine should be employed.

CHAPTER VII

THE CONDITIONS OF PERFECT OPERATION

General Care.—Gas-engines, as well as most machines in general, should be kept in perfect condition. Cleanliness, even in the case of parts of secondary importance, is indispensable. Unpainted and polished surfaces such as the shaft of the engine, the distributing cam-shafts, the levers, the connecting-rod and the like, should be kept in a condition equal to that when they were new. The absence of all traces of rust or corrosion in these parts affords sufficient evidence of the care taken of the invisible members such as the piston, the valves, ignition devices, and the like.

Lubrication.—The rubbing surfaces of a gas-engine should be regularly and perfectly lubricated. The absence of lost motion and backlash in the bearings, guides, and joints is of particular importance not only because of its influence on steady and silent running, but also on the power developed and on the consumption. As we have already seen in the chapter on lubrication, a special quality of oil should be employed for the lubrication of the cylinder. The feed of the lubricator supplying this most vital part of the engine is so regulated that it meets the actual requirements with the utmost nicety possible. In a subsequent chapter, in which faulty operation will be discussed, it will be shown how too much and too little oil may cause serious trouble.

Tightness of the Cylinder.—The amount of power developed depends principally on the degree of compression to which the explosive mixture is subjected. The economical operation of the engine depends in general upon perfect compression. It is, therefore, necessary to keep those parts in good order upon which the tightness of the cylinder depends. These parts are the piston, the valves, and their joints, and the ignition devices whether they be of the hot-tube or electrical variety. In order to prevent leakage at the piston, the rings should be protected from all wear. It is of the utmost importance that the surfaces both of the piston and of the cylinder, be highly polished so that binding cannot occur. In cleansing the cylinder, emery paper or abrasive powder should not be employed; for the slightest particle of abrasive between the surfaces in contact will surely cause leakage. The oil and dirt, which is turned black by friction and which may adhere to the piston rings, should be washed away with petroleum. Similarly the other parts of the cylinder should be cleaned to which burnt oil tends to adhere.

Valve-Regrinding.—The valves should be regularly ground. Even in special cases where they may show no trace of rapid wear they should be removed at least every month. In order to avoid any accident, care should be taken in adjusting the valves after the cap has been unbolted not to introduce a candle or a lighted match either in the valve-chambers or in the cylinder, without first closing the gas-cock. Furthermore, a few turns should be given to the engine, in order to drive out any explosive mixture that may still remain in the cylinder or the connected passages. The exhaust-valve, by reason of the high temperature to which the disk and the seat are subjected, should receive special attention. The valve should be ground on its seat every two or three months at least, depending upon the load of the engine.

Bearings and Crosshead.—The bushings of the engine shaft should always be held tightly in place. The looseness to which they are liable, particularly in gas-engines on account of the sharp explosions, tends to unscrew the nuts and to hasten the wear of the brass, which is the result of frequent tightening. The slightest play in the bearings of the engine-shaft as well as in the bearings of connecting-rods increases the sound that engines naturally produce.

Governor.—The governor should receive careful attention so far as its cleanliness is concerned; for if its operation is not easy it is apt to become "lazy" and to lose its sensitiveness. If the governor be of the ball type, or of the conical pendulum type operated by centrifugal force, it is well to lubricate each joint without excess of oil. In order to prevent the accumulation and the solidification of oil, the governor should be lubricated from time to time with petroleum. If the governor is actuated by inertia, which is the case in most engines of the hit-and-miss variety, it needs less care; still, it is advisable to keep the contact at which the thrust takes place well oiled.

The operation of any of these governors is usually controlled by the tension of a spring, or by a counterweight. In order to increase the speed of the engine, or in other words, to increase the number of admissions of gas in a given time, all that is usually necessary is to tighten up the spring, or to change the position of the counterweight. It should be possible to effect this adjustment while the engine is running in such a manner that the speed can be easily changed.

Joints.—In most well-built engines the caps of the valve-chests and other removable parts are secured "metal on metal" without interposing special joints. In other words, the surfaces are themselves sufficiently cohesive to insure perfect tightness. In engines which are not of this class, asbestos joints are very frequently employed, particularly at the exhaust-valve cap and the suction-valve.

In some engines, where for any reason it is necessary frequently to detach the caps, certain precautions should be taken to protect the joints so that they may not be exposed to deterioration whenever they are removed. For this purpose, they are first immersed in water in order to be softened, then dried and washed with olive or linseed oil on the side upon which they rest in the engine. On the cap side they are dusted with talcum or with graphite. Treated in this manner, the joint will adhere on one side and will be easily released on the other. Joints that are liable to come in contact with the gases in the explosion-chamber should be free from all projections toward the interior of the cylinder; for during compression these uncooled projections may become incandescent and may thus cause premature ignition. As a general rule when the cap is placed in position the joint should be retightened after a certain time, when the surfaces have become sufficiently heated. In order to tighten the joints the bolts and nuts should not be oiled; otherwise the removal of the cap becomes difficult.

Water Circulation.—In a previous chapter, the importance of the water circulation and the necessity of keeping the cylinder-jacket hot, have been sufficiently dwelt upon. As the cylinder tends to become hotter with an increase in the load, because of the greater frequency of explosions, it is advisable to regulate the flow of the water in order to prevent its becoming more than sufficient in quantity when the engine is lightly loaded; for under these conditions the cylinder will be cold and the explosive mixture will be badly utilized. A suitable temperature of 140 to 158 degrees F. is easily maintained by adjusting the circulation of the water. This can be accomplished by providing the water-inlet pipe leading to the cylinder with a cock which can be opened more or less, as may be necessary. The temperature of 140 to 158 degrees F. which has been mentioned may, at first blush, seem rather high, because it would be impossible to keep the hand on the outlet-pipe. The cylinder, however, will not become overheated so long as it is possible to hold the hand beneath the jacket near the water-inlet. This relates only to engines having a compression of 50 to 100 lbs. per square inch. For engines of higher compression, a lower running temperature will be safer. On this matter the instructions of the engine maker should be carried out.

Adjustment.—Gas-engines, at least those which are built by trustworthy firms, are always put to the brake test before they are sent from the shops, and are adjusted to meet the requirements of maximum efficiency. But since the nature and quality of gas necessarily vary with each city, it is evident that an engine adjusted to develop a certain horse-power with a gas of a certain richness, may not fulfil all expectations if it is fed with a gas less rich, less pure, hotter, and the like. The altitude also has some influence on the efficiency of the engine. As it increases, the density of the mixture diminishes; that is to say, for the same volume the engine is using a smaller amount. From this it follows that a gas-engine ought to be adjusted as a general rule on the spot where it is to be used.

The fulfilment of this condition is particularly important in the case of explosion-engines, because an advancement or retardation of only one-half a second in igniting the explosive mixture will cause a considerable loss in useful work. From this it would follow that gas-engines should be periodically inspected in order that they may operate with the highest efficiency and economy. As in the case of steam-engines, it is advisable to take indicator records which afford conclusive evidence of the perturbations to which every engine is subject after having run for some time.

Most gas-engine users either have no indicating instruments at their disposal or else are not sufficiently versed in their employment and the interpretation of their records to study perturbations by their means. For this reason the advice of experts should be sought,—men who understand the meaning of the diagrams taken and who are able by their means to effect a considerable saving in gas.

CHAPTER VIII

HOW TO START AN ENGINE—PRELIMINARY PRECAUTIONS

THE first step which is taken in starting an engine driven by street-gas is, naturally, the opening of the meter-cock and the valves between the meter and the engine. When the gas has reached the engine, the rubber bags will swell up and the anti-pulsator diaphragm will be forced out. The drain-cock of the gas-pipe is then opened. In order to ascertain whether the flow of gas is pure, a match is applied to the outlet of the cock. The flame is allowed to burn until it changes from its original blue color to a brilliant yellow.

If the hot-tube system of ignition be employed, the Bunsen burner is ignited, care being taken that the flame emerging from the tube is blue in color. If necessary the admission of air to the burner is regulated by the usual adjusting-sleeve. A white or smoky flame indicates an insufficient supply of air to the burner. A characteristic sooty odor is still other evidence of the same fact. Sometimes a white flame may be produced by the ignition of the gas at the opening of the adjusting-sleeve. A blue or greenish flame is that which has the highest temperature and is the one which should, therefore, be obtained. About five or ten minutes are required to heat up the tube, owing to the material of which it is made. When the proper temperature has been attained the tube becomes a dazzling cherry red in color. While the tube is being heated up, it is well to determine whether the engine is properly lubricated and all the cups and oil reservoirs are duly filled up. The cotton waste of the lubricators should be properly immersed, and the drip lubricators examined to determine whether they are supplying their normal quantity of oil.

The regulating-levers of the valves should be operated in order to ascertain whether the valves drop upon their seats as they should. The stem of the exhaust-valve should be lubricated with a few drops of petroleum.

If the ignition system employed be of the electric type, with batteries and coils, tests should be made to determine whether the current passes at the proper time on completing the circuit with the contact mounted on the intermediary shaft. This contact should produce the characteristic hum caused by the operation of the coil.

If a magneto be used in connection with the ignition apparatus, its inspection need not be undertaken whenever the engine is started, because it is not so likely to be deranged. Still, it is advisable, as in the case of ignition by induction-coils, to set in position the device which retards the production of the spark. This precaution is necessary in order to avoid a premature explosion, liable to cause a sharp backward revolution of the fly-wheel.

After the ignition apparatus and the lubricators have been thus inspected, the engine is adjusted with the piston at the starting position, which is generally indicated by a mark on the cam-shaft. The starting position corresponds with the explosion cycle and is generally at an angle of 40 to 60 degrees formed by the crank above the horizontal and toward the rear of the engine. The gas-cock is opened to the proper mark, usually shown on a small dial. If there be no mark, the cock is slowly opened in order that no premature explosion may be caused by an excess of gas.

The steps outlined in the foregoing are those which must be taken with all motors. Each system, however, necessitates peculiar precautions, which are usually given in detailed directions furnished by the builder.

As a general rule the engines are provided on their intermediary shafts with a "relief" or "half-compression" cam. By means of this cam the fly-wheel can be turned several times without the necessity of overcoming the resistance due to complete compression. Care should be taken, however, not to release the cam until the engine has reached a speed sufficient to overcome this resistance.

Engines of considerable size are commonly provided with an automatic starting appliance. In order to manipulate the parts of which this appliance is composed, the directions furnished by the manufacturer must be followed. Particularly is this true of automatic starters comprising a hand-pump by means of which an explosive mixture is compressed,—true because in the interests of safety great care must be taken.

The tightness and free operation of the valves or clacks which are intended to prevent back firing toward the pump should be made the subject of careful investigation. Otherwise, the piston of the pump is likely to receive a sudden shock when back firing occurs.

When the engine has been idle for several days, it is advisable, before starting, to give it several turns (without gas) in order to be sure that all its parts operate normally. The same precaution should be taken in starting an engine, if a first attempt has failed, in order to evacuate imperfect mixtures that may be left in the cylinder. Before this test is made, the gas-cock should, of course, be closed in order to prevent an untimely explosion. It is advisable in starting an engine not to bend the body over the ignition-tube, because the tube is likely to break and to scatter dangerous fragments.

Under no condition whatever should the fly-wheel be turned by placing the foot upon the spokes. All that should be done is to set it in motion by applying the hand to the rim.

Care During Operation.—When the engine has acquired its normal speed, the governor should be looked after in order that its free operation may be assured and that all possibility of racing may be prevented.

After the engine has been running normally for a time, the cocks of the water circulation system should be manipulated in order to adjust the supply of water to the work performed by the engine. In other words the cylinder should be kept hot, but not burning, as previously explained in the paragraph in which the water-jacket is discussed. The maintenance of a suitable temperature is extremely important so far as economy is concerned. All the bearings should be inspected in order that hot boxes may be obviated.

Stopping the Engine.—The steps to be taken in stopping the engine are the following:

1. Stopping the various machines driven by the engine,—a practice which is followed in the case of all motors;

2. Throwing out the driving-pulley of the engine itself, if there be one;

3. Closing the cock between the meter and the gas-bags in order to prevent the escape of gas and the useless stretching of the rubber of the bags or of the anti-pulsating devices;

4. Actuating the half-compression or relief cam as the motor slows down, in order to prevent the recoil due to the compression;

5. Closing the gas-admission cock;

6. Shutting off the supply of oil of free flowing lubricators, and lifting out the cotton from the others.

If the engine be used to drive a dynamo, particularly a dynamo provided with metal brushes, the precaution should be taken of lifting the brushes before the engine is stopped in order to prevent their injury by a return movement of the armature-shaft;

7. Shutting off the cooling-water cock if running water is used.

If the engine is exposed to great cold, the freezing of the water in the jacket is prevented while the engine is at rest, either by draining the jacket entirely, or by arranging a gas jet or a burner beneath the cylinder for the purpose of causing the water to circulate. If such a burner be used the cocks of the water supply pipe should, of course, be left open.

CHAPTER IX

PERTURBATIONS IN THE OPERATION OF ENGINES AND THEIR REMEDY

In this chapter will be discussed certain perturbations which affect the operations of gas-engines to a more marked degree than lack of care in their construction. In previous chapters defects in operation due to various causes have been dwelt upon, such as objectionable methods in the construction of an engine, ill-advised combination of parts, defects of installation, and the like; and an attempt has been made to determine in each case the conditions which must be fulfilled by the engine in order to secure efficiency and economy at a normal load.

Difficulties in Starting.—The preliminary precautions to be taken in starting an engine having been indicated, it is to be assumed that the advice given has been followed. Nevertheless various causes may prevent the starting of the engine.

Faulty Compression.—Defective compression, as a general rule, prevents the ignition of the explosive mixture. Whether or not the compression be imperfect can be ascertained by moving the piston back to the period corresponding with compression, in other words, that position in which all valves are closed.

If no resistance be encountered, it is evident that the air or the gaseous mixture is escaping from the cylinder by way of the admission-valve, the exhaust-valve, or the piston. The valves, ordinarily seated by springs, may remain open because their stems have become bound, or because some obstruction has dropped in between the disk and the seat. In a worn-out or badly kept engine the valves are likely to leak. If that be the case grinding is the only remedy. If a valve be clogged, which becomes sufficiently evident by manipulating the controlling levers, it is necessary simply to clean the stem and its guides in order to remove the caked oil which accumulates in time. If the engine be new, the binding of the valve-stems is often caused by insufficient play between the stems and their guides. Should this prove to be the case, the defect is remedied by rubbing the frictional surface of the stem with fine emery paper and by lubricating it with cylinder-oil. The exhaust-valve, however, should be lubricated only with petroleum.

It is not unlikely that the exhaust-valve may leak for two other reasons. In the first place, the tension of the spring which serves to return the valve may have lessened and may be insufficient to prevent the valve from being unseated during suction. Again, the screw or roller serving as a contact between the lever and the valve-stem, may not have sufficient play, so that the lengthening of the stem on account of its expansion may prevent the valve from falling back on its seat. The first-mentioned defect is remedied by renewing the spring, or by the provision of an additional spring or of a counterweight in order to prevent the stoppage of the motor. The second defect can be remedied by regulating the contact.

Leakage past the piston may be caused by the breaking of one or more rings, by wear or binding of the rings, or by wear or binding of the cylinder. The whistling caused by the air or the mixture as it passes back proves the existence of this fault.

Presence of Water in the Cylinder.—It may sometimes happen that water may find its way into the cylinder with the gas by reason of the bad arrangement of the piping. It may also happen that water may enter the cylinder through the water-jacket joint. Again, the presence of water in the cylinder may be due to condensation of the steam formed by the chemical union of the hydrogen of the gas and the oxygen of the air, which condensation is caused by the cool walls of the cylinder. The water may sometimes accumulate in the exhaust pipe and box, when they have been improperly drained, and may thus return to the cylinder. Whatever may be its cause, however, the presence of water in the cylinder impedes the starting of the engine, because the gases resulting from the explosion are almost spontaneously chilled, thereby diminishing the working pressure.

If electric ignition be employed, drops of water may be deposited between the contacts, thereby causing short circuits which prevent the passing of the spark.

If there be no drain-cock on the cylinder, the difficulty of starting the engine can be overcome only by ceaseless attempts to set it in motion. The leaky condition of a joint as well as the presence of a particle of gravel in the cylinder-casting, through which the water can pass from the jacket, is attested by the bubbling up of gas in the water-tank at the opening of the supply tube. These bubbles are caused by the passage of the gas through the jacket after the explosion. If such bubbles be detected, the cylinder should be renewed or the defect remedied. In order to obviate any danger, the stop-cocks of the water-jacket, which have already been described in a previous chapter, should be closed while the engine is idle.

Imperfect Ignition.—The difficulties encountered in starting an engine, and caused by imperfect ignition, vary in their nature with the character of the ignition system employed, whether that system, for example, be of the electric, or of the incandescent or hot tube type. Frequently it happens that in starting an engine a hot tube may break. If the tube be of porcelain the accident may usually be traced to improper fitting or to the presence of water in the cylinder. If the tube be of metal, its breaking is caused usually by a weakening of the metal through long use—an accident that occurs more often in starting the engine than in normal operation, because the explosions at starting are more violent, owing to the tendency of the supply-pipes to admit an excess of gas at the beginning.

A misfire arising from a faulty tube in starting may be caused by an obstruction or by leaks at the joints or in the body of the tube itself, thereby allowing a certain quantity of the mixture to escape before ignition. This defect in the tube is usually disclosed by a characteristic whistling sound.

A tube may leak either at the bottom or at the top. In the first case, starting is very difficult, because the part of the mixture compressed toward the tube will escape through the opening before it reaches the incandescent zone. In the second case, ignition may be

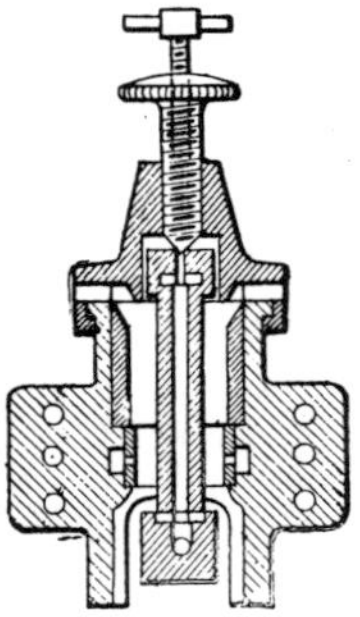

FIG. 74.

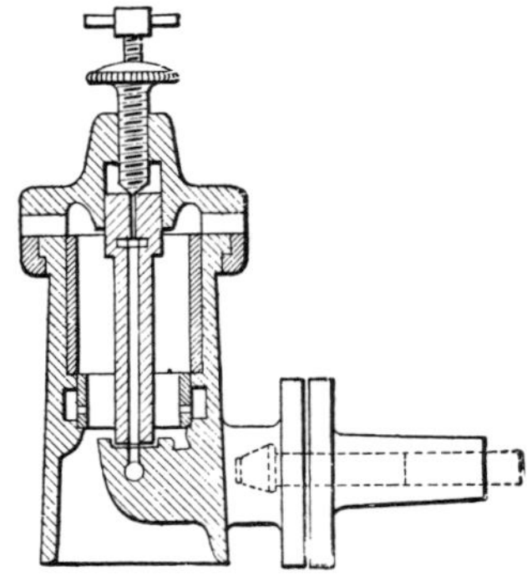

FIG. 75.

Ignition-tubes provided with needle valves to facilitate starting.

simply retarded to so marked an extent that a sufficient motive effect cannot be produced. An example of this retardation, artificially produced to facilitate the starting and to obviate premature explosions, is found in a system of ignition-tubes provided with a small cock or variable valve (Figs. 74 and 75).

The mere enumeration of defects caused by leakage is sufficient to indicate the remedy to be adopted. It may be well to recall in this connection the important part played by the ignition-valve. If it be leaky, or if its free operation be impeded, starting will always be difficult.

Electric Ignition by Battery or Magneto.—If the electric ignition apparatus, whatever may be the method by which the spark is produced, be imperfect in operation, the first step to be taken is to ascertain whether the spark is produced at the proper time, in other words, slightly after the dead center in the particular position given to the admission device at starting. If a coil and a battery be employed, it is advisable to remove the plug and to place it with its armature upon a well-polished metal surface to produce an electrical contact, preventing, however, the contact of the binding post with this metallic surface. The same method of inspection is adopted with the make-and-break apparatus of an electric magneto. In both cases it should be ascertained whether or not there is any short-circuiting. The contacts should be cleaned with a little benzine if they are covered with oil or caked grease.

If no spark is produced at the plug or at the make-and-break device it may be inferred that the wires are broken or that the generating apparatus is out of order. A careful examination will indicate what measures are to be taken to cure the defects.

Premature Ignition.—It has several times been stated that the moment of ignition of the gaseous mixture has a pronounced influence on the operation of gas-engines and upon their economy.

Premature ignition takes place when there is a violent shock at the moment when the piston leaps from the rear dead center to the end of the compression stroke. The violent effects produced are all the more harmful because they tend to overheat the interior of the engine and thereby to increase in intensity.

Premature ignition may be due to several causes. If a valveless hot tube be employed it may happen that the incandescent zone is too near the base. If the tube be provided with a valve, it very frequently happens that the valve leaks or that it opens too soon. In the case of electric ignition, the circuit may be completed before the proper time, because of faulty regulation. The suggestions made in the preceding chapters indicate the method of remedying these defects.

Faulty ignition may have its origin not only in the method of ignition employed, but also in excessive heating of the internal parts of the engine, caused by continual overloading or by inadequate circulation of water.

Passing to those cases of premature ignition of a special nature which are not due to any functional defect in the engine, but which are purely accidental in origin, such as the uncleanliness of the parts within the cylinder or the presence of some projecting part which becomes heated to incandescence during compression, it should first be stated that these ignitions, usually termed spontaneous, often occur well in advance of the end of the compression stroke. They are characterized by a more marked shock than that caused by ordinary premature ignition and usually result in bringing the engine to a complete stop in a very short time. These spontaneous explosions counteract to such an extent the impulse of the compression period, during which the piston is moving back, that they have a tendency to reverse the direction in which the engine is running. In such cases a careful inspection and a scrupulous cleaning of the cylinder and of the piston should be undertaken.

The bottom of the piston is particularly likely to retain grease which has become caked, and which is likely to become heated to incandescence and spontaneously to ignite the explosive mixture.

Untimely Detonations.—The sound produced by the explosions of a normally operating engine can hardly be heard in the engine-room. Untimely detonations are produced either at the exhaust, or in the suction apparatus, near the engine itself. These detonations are noisier than they are dangerous; still, they afford evidence of some fault in the operation which should be remedied.

Detonations produced at the exhaust are caused by the burning of a charge of the explosive mixture in the exhaust-pipe, which charge, for some reason, has not been ignited in the cylinder, and has been driven into the exhaust-pipe, where it catches fire on coming into contact with the incandescent gases discharged from the cylinder after the following explosion.

Detonations produced in the suction apparatus of the engine, which apparatus is either arranged in the base itself or in a separate chest, are often noisier than the foregoing. They are caused by the accidental backward flowing of the explosive mixture, and by its ignition outside of the cylinder. The accident may be traced to three causes:

1. The suction-valve of the mixture may not be tight and may leak during the period of compression, allowing a certain quantity of the mixture to pass into the suction-chest or into the frame. When the explosion takes place in the cylinder that part of the mixture which has passed back is ignited, as we have just seen, thereby producing a very loud deflagration. The obvious remedy consists in making the suction-valve tight by carefully grinding it.

2. It may happen that at the end of the exhaust stroke incandescent particles may remain in the cylinder, which particles may consist of caked oil or may be retained by poorly cooled projections. The result is that the mixture is prematurely ignited during the suction period.

3. The engine is so regulated, particularly in the case of English-built engines, as to effect what is technically called "scavenging" the products of combustion. In order to obtain this result, the mixture-valve is opened before the end of the exhaust stroke of the piston and the closing of the exhaust-valve. Owing to the inertia and the speed acquired by the products of combustion shot into the exhaust-pipe after explosion, a lowering of the pressure is produced in the cylinder toward the end of the stroke, causing the entrance of

air by the open admission-valve and consequently effecting the scavenging of the burnt gases, part of which would otherwise remain in the cylinder. It is evident that if a charge of the mixture has not been normally exploded, either because its constituents have not been mingled in the proper proportion, or because the ignition apparatus has missed fire, this charge at the moment of exhausting will pass out of the cylinder without any acquired speed, and will flow back in part at the end of the exhaust stroke past the prematurely opened admission-valve, thereby lodging in the air suction apparatus. Despite the suction which takes place immediately following the re-entrance of the gas into the cylinder, a certain quantity of the mixture is still confined in the suction-pipe and its branches, where it will catch fire at the end of the exhaust stroke after the opening of the mixture-valve.

In order to avoid these detonations it is necessary simply to see to it that the mixture is regularly ignited. This is accomplished by mixing the gas and air in proper proportions or by correcting the ignition time.

Retarded Explosions.—Retarded explosions considerably reduce the power which an engine should normally yield, and sensibly increase the consumption. They are due to three chief causes: (1), faulty ignition; (2), the poor quality of the mixture; (3), compression losses. The existence of the defect cannot be ascertained with any certainty without the use of an indicator or of some registering device which gives graphic records. Nevertheless, it is possible in some degree to detect retarded explosions, simply by observing whether there is a diminution in the power or an excessive consumption, despite the perfect operation and good condition of all the engine parts.

In order to remedy the defect it should be ascertained if the compression is good, if the supply of gas is normal, and if the conditions under which the mixture of air and gas is produced have not been changed. Lastly, the ignition apparatus is gradually adjusted to accelerate its operation until a point is reached when, after explosion, shocks are produced which indicate an excessive advance. The ignition apparatus is then adjusted to a point slightly ahead of the corresponding position. Recalling the descriptions already given of the various systems of ignition, the manner of regulating the moment of ignition in each case may be summarized as follows:

1. For the valveless incandescent tube, provided with a burner the position of which can be varied, ignition can be accelerated by bringing the burner nearer to the base. Retardation is effected by moving the burner away from the base.

2. In the case of the incandescent tube of the fixed burner type, the moment of ignition will depend upon the length of the tube. The retardation will be greater as the tube is shorter, and *vice versa*.

3. If the tube be provided with an ignition-valve, the time of ignition having been regulated by the maker, regulation need not be undertaken except if the valve-stem be worn or the controlling-cam be distorted.

If these defects should be noted, the imperfect parts should be repaired or renewed.

4. In electric igniters the controlling apparatus is generally provided with a regulating device which may be manipulated during the operation of the motor. If the manual adjustment of the regulating apparatus be unproductive of satisfactory results, it is advisable to ascertain whether the spark is being produced normally. Before the engine has come to a stop, one of the valve-casings is raised, and through the opening thus produced it is easily seen whether the spark is of sufficient strength, the engine in the meanwhile being turned by hand. Care should always be taken to purge the cylinder of the gas that it may contain, in order to prevent dangerous explosions. If the spark should prove to be too feeble, or if there be no spark at all, despite the fact that every part of the mechanism is properly adjusted, it may be inferred that the fault lies with the current and is caused by

1. Imperfect contact with the binding-posts, with the conducting wire, or with the contact-breaking members;

2. A short circuit in one of the dismembered pieces;

3. The presence of a layer of oil or of caked grease forming an insulator, injurious to induction, between the armature and the magnets;

4. A deposit of oil or moisture on the contact-breaking parts;

5. The exhaustion of the magnets, which, however, occurs only after several years of use, except when the

magneto has been subjected for a long time to a high temperature.

The mere discovery of any of these defects sufficiently indicates the means to be adopted in remedying them.

Lost Motion in Moving Parts.—Lost motion of the moving parts is due to structural errors. Its cause is to be found in the insufficient size of the frictional bearing surfaces, and improper proportioning of shafts, pins, and the like. The result is a premature wear which cannot be remedied. Imperfect adjustment, lack of care, and bad lubrication, may also hasten the wear of certain parts. This wear is manifested in shocks, occurring during the operation of the engine,—shocks which are particularly noticeable at the moment of explosion.

Besides the inconveniences mentioned, wearing of the gears and of the moving parts leads to derangement of the power-transmitting members.

So far as the admission and exhaust valves are concerned, the wearing of the cams, rollers, and lever-pivots is evidenced by a retardation in the opening of these valves and an acceleration in their closing.

The ignition, whatever may be the system employed, is affected by lost motion and is retarded. The engine appreciably loses in power, and its consumption becomes excessive.

Overheated Bearings.—Apart from the imperfect adjustment of a member, it may happen that the bushings of the main bearings of the ends of the connecting-rod, and of the piston-pivot, may become heated because of excessive play, or of too much tightening, or of a lack of oil, or of the employment of oil of bad quality. The overheating may lead to the binding of frictional surfaces and even to the fusion of bushings if they be lined with anti-friction metal. In order to avoid the overheating of parts, it is advisable, while the engine is running, to touch them from time to time with the back of the hand. As soon as the slightest overheating is felt, the temperature may be lowered often by liberal oiling. If this be inadequate and if for special reasons it is impossible to stop the engine, the overheated part may be cooled by spraying it with soapy water.

If the overheating has not been detected or reduced in time, a characteristic odor of burnt oil will be perceived, accompanied by smoke. The part overheated will then have attained a temperature so high that it cannot be touched with the hand. Should this occur, it is inadvisable to employ oil, because it would immediately burn up and would only aggravate the conditions. Cotton waste should be carefully applied to the overheated member, and gradual spraying with soapy water begun.

In special cases where the lubricating openings or channels are not likely to be obstructed, a little flowers of sulphur may be added to the oil, if this be very fluid. Castor oil may also be successfully employed.

If the binding of the rubbing surfaces should prevent the reduction of the overheated member's temperature, the engine must necessarily be stopped, and the parts affected detached. All causes of binding are removed by means of a steel scraper. The surfaces of the bushings and of the shaft which they receive are smoothed with a soft file and then polished with fine emery paper. Before the parts are replaced, the precaution of ascertaining whether they touch at all points should be taken. Careful inspection and copious lubrication should, of course, be undertaken when the engine is again started.

Overheating of the Cylinder.—The overheating of the cylinder may be due to a complete lack of water in the jacket or to an accidental diminution in the quantity of water supplied. If this discovery is made too late, and if the cylinder has reached a very high temperature, the circulation of the water should not be suddenly re-established, because of the liability of breaking the casting. It is best to stop the engine and to restore the parts to their normal condition.

It is well to recall at this point that if the calcareous incrustation of the water-jacket or the branch pipes should hinder the free circulation of water, cleaning is, of course, necessary. The jacket may be washed several times with a twenty per cent. solution of hydrochloric acid. After this treatment the jacket should, of course, be rinsed with fresh water before the piping of the water-circulating apparatus is again connected.

Overheating of the Piston.—If the overheating of the piston is not due to faulty adjustment, it may be caused by lack of oil or to the employment of a lubricant not suitable for the purpose. In a previous chapter the importance of using a special oil for cylinder lubrication has been insisted upon. The overheating of the piston can also result from that of the piston-pin. Should this be the case it is advisable to stop the engine, to ascertain the condition and the degree of lubrication of this member and its bearing. Overheating of the piston is manifested by an increase of the temperature of the cylinder at the forward end. If this overheating be not checked, binding of the piston in the cylinder is likely to result.

Smoke Arising from the Cylinder.—This is generally a sign either of overheating, which causes the oil to evaporate, or of an abnormal passage of gas, caused by the explosion. Abnormal passage of gas may result from wear or from distortion of the cylinder, or from wear or breakage of the piston-rings. The result is always the overheating of the cylinder and a reduction in compression and power.

If the engine is well kept and shows no sign of wear, leakage may be caused simply by the fouling of the piston-rings, which then adhere in their grooves and have but insufficient play. This defect is obviated by cleaning the rings in the manner explained in Chapter VII.

Lubrication is faulty when the quantity of lubricant supplied is either insufficient or too abundant, or when the oils employed are of bad quality. It has already been shown that insufficient lubrication and the utilization of bad oils leads to the overheating of the moving parts.

Insufficient lubrication may be caused by imperfect operation of the lubricators, or, particularly during cold weather, by too great a viscosity or congelation of the oil. If a lubricator be imperfect in its operation, the condition of its regulating mechanism should be ascertained, if it has any, and an examination made to discover any obstruction in the oil-ducts. Such obstructions are very likely to occur in new devices which have been packed in cotton waste or excelsior, with the result that the particles of the packing material often find their way into openings.

An oil may be bad in quality because of its very nature, or because of the presence of foreign bodies. In either case an oil of better quality should be substituted.

The freezing of oil by intense cold may be retarded by the addition of ordinary petroleum to the amount of 10 to 20 per cent.

An excess of oil in the bearings results simply in an unnecessary waste of lubricant, and the splashing of oil on the engine and about the room. If too much oil be used in the cylinder, grave consequences may be the result; for a certain quantity of the oil is likely to accumulate within the cylinder, where it burns and forms a caky mass that may be heated to incandescence and prematurely ignite the explosive mixture. Especially in producer-gas engines is an excess of cylinder-lubricant likely to cause such accidents. Indeed, the temperature of explosion not being as high as in street-gas engines, the excess oil cannot be so readily removed with certainty by evaporation or combustion. On the other hand, the compression of the mixture being generally higher, premature ignition is very likely to occur.

Back Pressure to the Exhaust.—How the pipes and chests for the exhaust should be arranged in order not to exert a harmful influence on the motor has already been explained. Even if the directions given have been followed, however, the exhaust may not operate properly from accidental causes. Among these causes may be mentioned obstructions in the form of foreign bodies, such as particles of rust, which drop from the interior of the pipes after the engine has been running for some time and which, accumulating at any place in the pipe, are likely to clog the passage. Furthermore, the products of combustion may contain atomized cylinder oil which finds its way into the exhaust-pipe. This oil condenses on the walls of the elbows and bends of the pipe in a deposit which, as it carbonizes, is converted into a hard cake and which reduces the cross-section of the passage, thereby constituting a true obstacle to the free exhaust of the gases.

These various defects are manifested in a loss in engine power as well as in an abnormal elevation of the temperature of the parts surrounding the exhaust opening.

Sudden Stops.—Sudden stops are occasioned by faulty operation of the engine, and by imperfect fuel supply. Among the first class the chief causes to be mentioned are the following:

1. Overheating, which has already been discussed and which may block a moving part.

2. Defective ignition.

3. Binding of the admission-valve or of the exhaust-valve, preventing respectively suction or compression.

4. The breaking or derangement of a member of the distributing mechanism.

5. A weakening of the exhaust-valve spring, so that the valve is opened by the suction of fresh quantities of mixture.

These faults are due to carelessness and improper inspection of the engine.

So far as the fuel supply of the engine is concerned, the causes of stoppage will vary if street-gas or producer-gas be employed. In the former case the difficulty may be occasioned by the improper operation of the meter, by the formation of a water-pocket in the piping, by the binding of an anti-pulsator valve, by the derangement of a pressure-regulator, or by a sudden change in the gas pressure when no pressure-regulator is employed. If producer-gas be used, stoppages may be occasioned by a sudden change in the quality, quantity, or temperature of the gas. These defects will be examined in detail in the chapter on Gas-Producers.

CHAPTER X

PRODUCER-GAS ENGINES

THUS far only street-gas or illuminating-gas engines have been discussed. If the engine employed be small —10 to 15 horse-power, for instance—street-gas is a fuel, the richness, purity and facility of employment of which offsets its comparatively high cost. But the constantly increasing necessity of generating power cheaply has led to the employment of special gases which are easily and cheaply generated. Such are the following:

Blast-furnace gases,
Coke-oven gases,
Fuel-gas proper,
Mond gas,
Mixed gas,
Water-gas,
Wood-gas.

The practical advantages resulting from the utilization of these gases in generating power were hardly known until within the last few years. The many uses to which these gases have been applied in Europe since 1900 have definitely proved the industrial value of producer-gas engines in general.

The steps which have led to this gradually increasing use of producer-gas have been learnedly discussed and commented upon in the instructive works and publications of Aimé Witz, Professor in the Faculty of Sciences of Lille, in those of Dugald Clerk, of London, F. Grover, of Leeds, and Otto Güldner, of Munich, and in those of the American authors, Goldingham, Hiscox, Hutton, Parsell and Weed, etc. The new tendencies in the construction of large engines may be regarded as an interesting verification of the forecasts of these men—forecasts which coincide with the opinion long held by the author. Aimé Witz has always been an advocate of high pressures and of increased piston speed. English builders who made experiments in this direction conceded the beneficial results obtained; but while they increased the original pressure of 28 to 43 pounds per square inch employed five or six years ago to the pressure of 85 to 100 pounds per square inch nowadays advocated, the Germans, for the most part, have adopted, at least in producer-gas engines, pressures of 114 to 170 pounds per square inch and more.

High Compression.—In actual practice, the problem of high pressures is apparently very difficult of solution, and many of the best firms still seem to cling to old ideas. The reason for their course is, perhaps, to be found in the fact that certain experiments which they made in raising the pressures resulted in discouraging accidents. The explosion-chambers became overheated; valves were distorted; and premature ignition occurred. Because the principle underlying high pressures was improperly applied, the results obtained were poor.

High pressures cannot be used with impunity in cylinders not especially designed for their employment; and this is the case with most engines of the older type, among which may be included most engines of English, French, and particularly of American construction. In American engines notably, the explosion-chamber, the cylinder and its jacket, are generally cast in one piece, so that it is very difficult to allow for the free expansion of certain members with the high and unequal temperatures to which they are subjected (Fig. 22).

Some builders have attempted to use high pressures without concerning themselves in the least with a modification of the explosive mixture. The result has been that, owing to the richness of the mixture, the explosive pressure was increased to a point far beyond that for which the parts were designed. Sudden starts and stops in operation, overheating of the parts, and even breaking of crank-shafts, were the results. The engines had gained somewhat in power, but no progress had been made in economy of consumption, although this was the very purpose of increasing the compression.

High pressures render it possible to employ poor mixtures and still insure ignition. A quality of street-gas, for example, which yields one horse-power per hour with 17.5 cubic feet and a mixture of 1 part gas and 8 of air compressed to 78 pounds per square inch, will give the same power as 14 cubic feet of the same gas mixed with 12 parts of air and compressed to 171 pounds per square inch.

"Scavenging" of the cylinder, a practice which engineers of modern ideas seem to consider of much importance, is better effected with high pressures, for the simple reason that the explosion-chamber, at the end of the return stroke, contains considerably less burnt gases when its volume is smaller in proportion to that of the cylinder.

In impoverishing the mixture to meet the needs of high pressures, the explosive power is not increased and in practice hardly exceeds 365 to 427 pounds per square

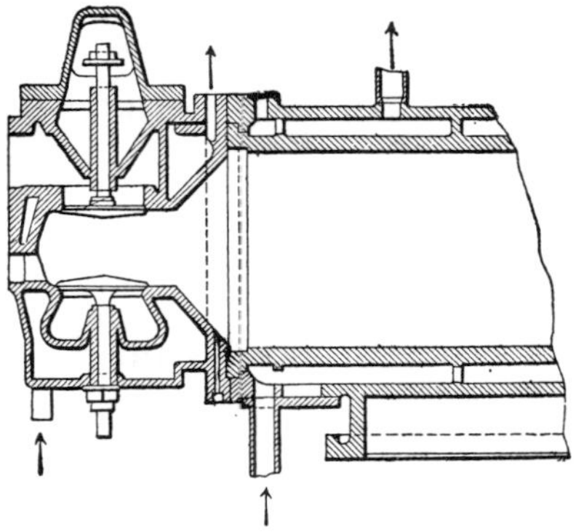

FIG. 76.—Method of cooling the cylinder-head.

inch. With the higher pressures thus obtained there is consequently no reason for subjecting the moving parts to greater forces.

Cooling.—The increase in temperature of the cylinder-head and of the valves, due wholly to high compression, is perfectly counteracted by an arrangement which most designers seem to prefer, and which, as shown in the accompanying diagram (Fig. 76), consists in placing the mixture and exhaust-valves in a passage forming a kind of antechamber completely surrounded by water. The immediate vicinity of this water assures the perfect and equal cooling of the valve-seats. This arrangement, while it renders it possible to reduce the size of the explosion-chamber to a minimum, has the additional mechanical advantage of enabling the builder to bore the seats and valve-guides with the same tool, since they are all mounted on the same line. From the standpoint of efficiency, the design has the advantage of permitting the introduction of the explosive mixture without overheating it as it passes through the admission-valve, which obtains all the benefit of the cooling of the cylinder-head, literally surrounded as it is by water.

In large engines the cooling effect is even heightened by separately supplying the jackets of the cylinder-head and of the cylinder. In engines of less power the top of the cylinder-head jacket is placed in communication with that of the cylinder, so that the coldest water enters at the base of the head and, after having there been heated, passes around the cylinder in order finally to emerge at the top toward the center. The water having been thus methodically circulated, the useful effect and regularity of the cooling process is increased.

Notwithstanding the care which is devoted to water circulation, it is advisable to run the producer-gas engine "colder" than the older street-gas types, in which the more economic speed is that at which the water emerges from the jacket at about a temperature of 104 degrees F. It would seem advisable to meet the requirements of piston lubrication by reducing to a minimum the quantity of heat withdrawn by the circulating water. Indeed, the personal experiments of the author bear out this principle.

For street-gas engines, however, the cylinders should be worked at the highest possible temperature consistent with the requirements of lubrication. It should not be forgotten that, in large engines fed with producer-gas, economy of consumption is a secondary consideration, because of the low quantity of fuel required. The cost, moreover, may well be sacrificed to that steadiness of operation which is of such great importance in large engines furnishing the power of factories; for in such engines sudden stops seriously affect the work to be performed. For this reason engine builders have been led to the construction of motors provided with very effective cooling apparatus. Since the circulation of the water around the explosion-chamber and the cylinder is not sufficient to counteract the rise of temperature, it has become the practice to cool separately each part likely to be subjected to heat. The seats of the exhaust-valves, the valves themselves, the piston, and sometimes the piston-rod, have been provided with water-jackets.

Premature Ignition.—Returning to the causes of the discouragements encountered by some designers who endeavored to use high pressures, it has already been mentioned that premature ignition of the explosive mixture in cylinders not suited for high pressures is one reason for the bad results obtained. An explanation of these results is to be found in the high theoretical temperature corresponding with great pressures and in the quantity of heat which must be absorbed by the walls of the explosion-chamber. These two circumstances are in themselves sufficient to produce spontaneous ignition of excessively rich mixtures, compressed in an overheated chamber unprovided with a sufficient circulation of water. A third cause of premature ignition may also be found in the old system of ignition which, in most English engines, consists of a metallic or porcelain tube, the interior of which communicates with the explosion-chamber, an exterior flame being employed to heat the tube to incandescence. In tubes of this type which are not provided with a special ignition-valve, the time of ignition is dependent only on the moment when the explosive mixture, driven into the tube, comes into contact, at the end of the compression stroke, with the incandescent zone, thereby causing the ignition. This very empirical method leads either to an acceleration or retardation of the ignition, depending upon the temperature of the tube, the position of the red-hot zone, its dimensions, and the temperature of the mixture, which is determined by the load of the engine. Although this system, the only merit of which is its simplicity, may meet the requirements of small engines, there is not the slightest doubt that it is quite inapplicable to those of more than 20 to 25 horse-power, for in such engines greater certainty in operation is demanded. Even if only the more improved of the two types of hot-tube ignition be considered, with or without valves, it must still be held that they are inapplicable to high compression engines. The ignition-valve is the part which suffers most from the high temperature to which it is subjected. Its immediate proximity to the incandescent tube, and its contact with the burning gas when it flares up, render it almost impossible to employ any cooling arrangement. Although with the exercise of great care it may work satisfactorily in engines of normal pressure, it is evident that it cannot meet the requirements of high-pressure engines, because the temperature of the compressed mixture is such that the charge is certain to catch fire by mere contact with the overheated valve. In industrial engines of small size, premature ignition has little, if any, effect except upon silent operation and economic consumption. This does not hold true, however, of large engines. Besides the inconveniences mentioned, there is also the danger of breaking the cranks or other moving parts. The inertia of these members is a matter of some concern, because of their weight and of the linear speed which they attain in large engines. Some idea of this may be obtained when it is considered that in a producer-gas or blast-furnace-gas engine having a piston diameter of 24 inches and an explosive pressure of 299 pounds per square inch, the force exerted at the moment of explosion is about 132,000 pounds. Naturally, engine builders have adopted the most certain means of avoiding premature ignition and its grave consequences.

The method of ignition which at present seems to be preferred to any other for producer-gas is that employing a break-spark obtained with the magneto apparatus previously described. Some builders of large engines, particularly desirous of assuring steadiness of running, have provided the explosion-chamber with two independent igniters. It may be that they have adopted this arrangement largely for the purpose of avoiding the inconveniences resulting from a failure of one of the igniters, rather than for the purpose of igniting the mixture in several places so as to obtain a more uniform ignition and one better suited for the propagation of the flame.

The Governing of Engines.—Various methods have been adopted for the purpose of varying the motive power of an engine between no load and full load, still preserving, however, a constant speed of rotation. These methods consist in changing either the quantity or the quality of the mixture admitted into the cylinder. Thus it may happen that an engine may be supplied:

1. With a mixture constant in quality and in quantity;

2. With a mixture variable in quality and constant in quantity;

3. With a mixture constant in quality and variable in quantity.

1. *Mixture Constant in Quality and Quantity.*—This method implies the use of the hit-and-miss system of admission, in which the number of admissions and explosions varies, while the value or the composition of each admitted charge remains as constant as the compression itself (Fig. 34). This system has already been

referred to and its simplicity fully set forth. By its use a comparatively low consumption is obtained, even when the engine is not running at full load. On the other hand, it has the disadvantage of necessitating the employment of heavy fly-wheel to preserve cyclic regularity.

2. *Mixture Variable in Quality and Constant in Quantity.*—The governing system most commonly employed to obtain a mixture variable in quality and constant quantity is based upon the control of the gas-admission valve by means of a cam having a conical longitudinal section, as shown in Fig. 35. This cam, commonly called a "conical cam," is connected with a lever actuated from the governor. As the lever swings under the action of the governor, the cam is shifted along the half-speed shaft of the engine. The result is that the gas-admission valve is opened for a longer or shorter period.

In another system a cylindrical valve is mounted between the chamber in which the mixture is formed and the gas-supply pipe, the valve being carried on the same stem as the mixture-valve itself. The cylindrical valve is displaced by the governor so as to vary the quantity of gas drawn in with relation to the quantity of air.

When the engines are fed with producer-gas the parts which have just been described should be frequently inspected and cleaned; for they are only too easily fouled.

Engines thus governed should be run at high pressure so as to insure the ignition of the producer-gas mixtures formed when the position of the cam corresponds with the minimum opening of the gas-valve. Powerful governors should be employed, capable of overcoming the resistance offered by the cylindrical valve or the cam.

It may often happen that variations in the load of the engine render it necessary to actuate the air valve, so as to obtain a mixture which will be ignited and exploded under the best possible conditions.

3. *Mixture Constant in Quality and Variable in Quantity.*—In supplying an engine with a mixture constant in quality and variable in quantity, the compression does not remain constant. The quantity of mixture drawn in by the cylinder may even be so far reduced that the pressure drops below the point at which ignition takes place. For that reason engines of this type should be run at high pressures.

The variation of the quantity of mixture may be effected in various ways. The simplest arrangement consists in mounting a butterfly-valve in the mixture pipe, which valve is controlled by the governor and throttles the passage to a greater or lesser degree. A very striking solution of the problem consists in varying the opening of the mixture-valve itself. To attain this end the valve is moved by levers. The point of application of one of these levers is displaced under the action of the governor so as to vary the travel of the valve within predetermined limits. Under these conditions a mixture of constant homogeneity is introduced into the cylinder, so proportioned as to insure ignition even at low pressures.

In recent experiments conducted by the author it was proved that with this governing system ignition still takes place even though the pressure has dropped to

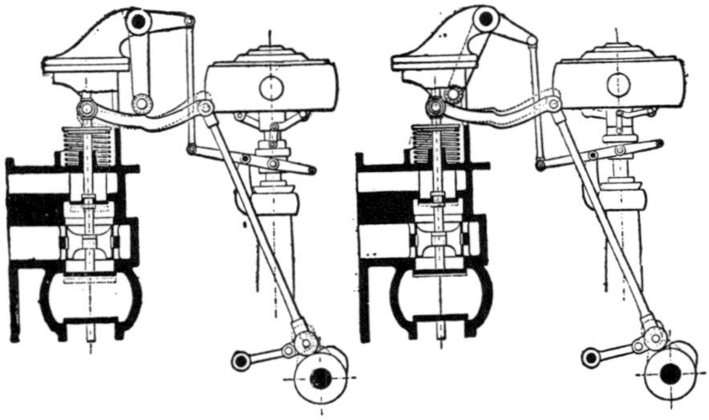

FIG. 76a.—Governing system for producer-gas engines.

43 pounds per square inch. This system has the merit of rendering it possible to employ ordinary governors of moderate size, since the resistance to be overcome at the point of application of the lever is comparatively small. In the accompanying illustration the Otto Deutz system is illustrated.

CHAPTER XI

PRODUCER-GAS

It may here be not amiss to point out the differences between illuminating gas and those gases which are called in English "producer" gases, and in French "poor" gases, because of their low calorific value.

Street-Gas.—This gas, the composition of which varies with different localities, has a calorific value, which is a function of its composition, and which varies from 5,000 to 5,600 calories per cubic meter (19,841 to 24,896 B. T. U. per 35.31 cubic feet) measured at constant pressure and corrected to 0 degrees C. (32 degrees F.) at a pressure of 760 millimeters (29.9 inches of mercury, or atmospheric pressure), not including the latent heat of the water of condensation. The following table gives the average volumetric composition of illuminating gas in various cities:

	Cities.				
	London.	Manchester.	New York.	Paris.	Berlin.
Hydrogen	48	46	40	52	50
Carbon monoxide	4	7	4	6	9
Methane	38	35	37	32	33
Various hydrocarbons	4	6	7	6	5
Carbon dioxide	..	4	3	..	2
Nitrogen	5	2	8	4	1
Oxygen	1	..	1	..	..
	100	100	100	100	100

Furthermore, these constituents vary within certain limits. This is also true of the calorific value. Experiments made by the author have demonstrated that in the same place at an interval of a few hours, variations of approximately ten per cent. occur.

Composition of Producer-Gases. — The average chemical composition of producer-gases varies with the conditions under which they are generated and the nature of the fuel. The following are the proportions of its constituents expressed volumetrically:

	Gas.					
	Blast Furnace.	Producer.	Mond.	Mixed (Fichet).	Water (Strache).	Wood (Riché).
Nitrogen and oxygen....	60	59	42	50	5	1
Carbon monoxide......	24	25	11	20	40	29
Carbon dioxide........	12	5	16	7	4	11
Hydrocarbons.........	2	2	2	3	1	15
Hydrogen............	2	9	29	20	50	44
	100	100	100	100	100	100
Calorific value in calories.	950	1,100	1,400	1,300	2,400	2,960
Average weight of a cubic meter in kilos.......	1.30	1.10	1.02	1.05	0.680	0.824
Or of a cubic foot in pounds............	0.008	0.007	0.006	0.0068	0.0042	0.0051

Blast-furnace gas has been used for generating power by means of gas-engines for about ten years. At the present time it is used in engines of very high power, a discussion of which engines more properly belongs to a work on metallurgy, and has no place, therefore, in a manual such as this.

Producer-gas, in the true sense of the term, is generated in special apparatus either under pressure or by suction in a manner to be described in the following chapters.

Mond gas is produced in generators of the blowing or pressure type from bituminous coal, necessitating the employment of special purifiers and permitting the collection of the by-products of the fractional distillation of the coal. Mond gas plants are, therefore, rather complicated and can be advantageously utilized only for large engines. More exhaustive information can be obtained from the descriptions published by the builders of Mond gas generators.

Mixed gas is generated in apparatus arranged so that the retort is kept at a high temperature, thereby producing a gas richer in hydrogen than that made by producers. It should be observed that in practice the generators at present used yield a producer-gas, the calorific value of which fluctuates between 1,000 and 1,400 calories per cubic meter (3,968 to 5,158 B. T. U. per 35.31 cubic feet); and the composition varies accordingly, in the manner that has already been indicated in the tables for producer-gas and mixed gas. There is no necessity, therefore, for drawing a distinction between these two qualities of gas.

Water-gas should theoretically be composed of 50 per cent. carbon monoxide and 50 per cent. hydrogen, resulting from the decomposition of steam by incandescent coal. In practice, however, it contains a little nitrogen and carbon dioxide. The gas is obtained from generators in which air is alternately blown in to fan the fire and then steam to produce gas. Water-gas is employed in soldering on account of its reducing properties and of the high temperature of its flame. The great quantity of carbon monoxide which it contains renders it very poisonous and exceedingly dangerous, because it is generated under pressure. From the economical standpoint, its generation is more expensive than that of producer-gas, for which reason its employment in gas-engines is hardly of much value.

Wood-gas, the composition of which has already been given, is generated in apparatus of the Riché type, the principle of which consists in heating a cast retort charged with any kind of fuel, namely wood, and vertically mounted on a masonry base.

This apparatus should be of particular interest to the proprietors of sawmills, furniture factories, and the like, since it offers a means of using the waste products of their plants.

The relatively high proportion of carbon monoxide in producer-gas is objectionable from a hygienic standpoint, so much so, indeed, that it has attracted the attention of manufacturers. Carbon monoxide, the specific gravity of which is 0.967, is a gas peculiarly poisonous and dangerous. It cannot be breathed without baneful effects, and is even more dangerous than carbonic-acid gas, which eventually causes asphyxiation by reducing the quantity of oxygen in the air. For this reason, it is necessary to take the utmost precaution in efficiently and continuously ventilating the rooms in which the gas-generators and their accessories are installed. This suggestion should be followed, above all, when the apparatus in question are installed in cellars and basements. As a further precaution, where the plant is rather large a workman should not be allowed to enter the generator room alone.

Blowing-generators, or those in which the gas is produced under pressure, are more dangerous than suction-generators. In the former a leaky joint may cause the vitiation of the surrounding air as the producer-gas escapes; in the suction apparatus the same fault simply causes more air to be drawn in.

Dr. Melotte recommends the following procedure in cases of carbon monoxide asphyxiation:

Carbon Monoxide Asphyxiation

Cases of poisoning by carbon monoxide are both frequent and dangerous. The gas is extremely poisonous, and all the more dangerous because it is odorless, colorless and tasteless. When it comes into contact with the blood, it forms a combination so stable that it is reacted upon by the oxygen of the air only with difficulty. It follows, therefore, that with each respiration of air charged with carbon monoxide, a certain quantity of blood is poisoned. In consequence of this, there is a possibility of poisoning in open air.

Symptoms.—The symptoms observed will vary with the manner in which the blood has been poisoned. There are two ways in which this poisoning can occur. The one depends upon whether the atmosphere contains an excess of carbon monoxide; the other whether the air breathed contains only traces of the gas.

Gradual, Rapid Asphyxiation.—At first a vague sickness is felt, rapidly followed by violent headaches, vertigo, anxiety, oppression, dimness of vision, beating of the pulse at the temples, hallucinations, and an irresistible desire to sleep. If at this stage the patient has a sufficient idea of danger to prompt him to open a window or door, he will escape death.

In the second stage, the victim's legs are paralyzed, but he can still move his arms and his head. The mind still preserves its clearness, and in a measure assists the further process of asphyxiation because of its impotency. Then follow coma and death.

Slow, Chronic Asphyxiation.—Slow, chronic asphyxiation is not infrequent. Its symptoms are often difficult to detect. Poisoning is manifested by weakness, cephalalgia, vomiting, pallor, general anemia, lassitude, and local paralysis. If any of these symptoms appear in the men who work in the vicinity of the producers, immediate steps should be taken to prevent the possibility of carbon monoxide asphyxiation.

First Aid in Cases of Carbon Monoxide Poisoning

It has already been stated that the oxygen of the air has no oxidizing effect upon blood contaminated by carbon monoxide. Only a liberal current of pure oxygen can oxidize the combination formed and render hematosis possible. This liberal current can be obtained from an oxygen tank of the portable variety, provided with a tube carrying at its free end a mask which is held over the mouth and the nostrils. The absorption of gas takes place by artificial respiration, which is effected in several ways. The most practical of these are the Sylvester and Pacini methods.

Sylvester Method.—The patient is laid on his back. His arms are raised over his head and then brought back on each side of the body. This operation is repeated fifteen times per minute approximately. The method is very frequently employed and is excellent in its results.

The Pacini Method.—Four fingers are placed in the pit of the arm, with the thumb on the shoulder. The shoulder is then alternately raised and lowered, producing a marked expansion of the chest. This method is the more effective of the two. The movements described are repeated fifteen to twenty times each minute very rhythmically.

One or the other of these two methods of treatment should be immediately applied in serious cases. Certain preliminary precautions should be taken in all cases, however. The patient should be carried to a well-ventilated and moderately heated room, stripped of his clothes, and warmed by water-bottles and heated linen. Reflex action should be excited, the peripheral nervous system stimulated in order to contract the heart and the respiratory muscles, and the precordial region cauterized. In addition to this treatment, the region of the diaphragm should be rubbed and pinched, the skin rubbed, cold showers given, flagellations administered, urtications (whipping with nettles) undertaken, the skin and the mucous membranes excited, the mucous membrane of the nose and of the pharynx titillated with a feather dipped in ammonia, alcohol, vinegar, or lemon juice.

Rhythmic traction of the tongue is effective when carried out as follows: The tongue is seized with a forceps and kept extended by means of a coarse thread. It is then pulled out from the mouth sharply and allowed to reenter after each traction. These movements should be rhythmic and should be repeated fifteen to twenty times a minute.

All these efforts should be continued for several hours. When the patient has finally been revived, he should be placed in a warm bed. Stimulants such as wine, coffee, and the like should be administered. If the head should be congested, local blood-letting should be resorted to and four or six leeches applied behind the ears. It should be borne in mind that the various steps enumerated are to be taken pending the arrival of a physician.

Impurities of the Gases

Most of the coal used in generating producer-gas contains sulphur. Sulphuretted hydrogen is thus produced, which mixes with the gas and imparts to it its characteristic odor. In some gas-generators, purifiers are employed in which sawdust mixed with iron salts is utilized, with the result that a combination is formed with the sulphuretted hydrogen, thereby removing it from the producer-gas. In other forms of generators a more summary method of purification is adopted, so that traces of sulphuretted hydrogen still remain. Since this gas attacks copper, the employment of this metal is not advisable for the following apparatus: Generator (openings, cock for testing the gas); piping (gas-pressure cocks, drain and pet cocks); engine (gas-admission cock, lubricating joint in the cylinder, valves and cocks of the compressed-air starting-pipe).

The distillation of coal in generators results in the formation of ammonia gas. This also has a corrosive action on copper and its alloys; but owing to its great solubility, it is eliminated by the waters of the "scrubber" and does not reach the engine.

Production and Consumption

The quantity of gas produced in most generators varies from 6.4 to 8.2 pounds per cubic foot of raw coal burnt in the generator. The engine consumes per horse-power per hour 70 to 115 cubic feet of gas, depending upon its richness.

CHAPTER XII

PRESSURE GAS-PRODUCERS

As we have already seen, producer-gas as a fuel for engines may be generated in two kinds of apparatus, the one operating under pressure, and the other by suction.

Dowson Gas-Producers.—The first pressure-generators were introduced by Dowson of London and necessitated installations of quite a complicated nature. Later improvements made by the designers contributed much to the general employment of their system. Many installations varying from 50 to 100 horse-power and more may be found in the United Kingdom, all of them made by Dowson. Indeed, for a long time the name of Dowson was coupled with producer-gas itself. The Dowson system necessitates the utilization of anthracite or of comparatively hard coal, such as that mined in Wales and Pennsylvania. Owing to the necessity of employing this special quality of coal the Dowson system and the systems that sprang from it were burdened with cooling, washing, and purifying apparatus, which complicated the installations to such an extent that they resembled gas works. The generator that took the place of the retort was fed with air and steam, blown in under pressure, necessitating the employment of a boiler. Furthermore, the production of the gas under pressure necessitated the use of a gasometer for its collection before it was supplied to the engine-cylinder. Such installations were evidently

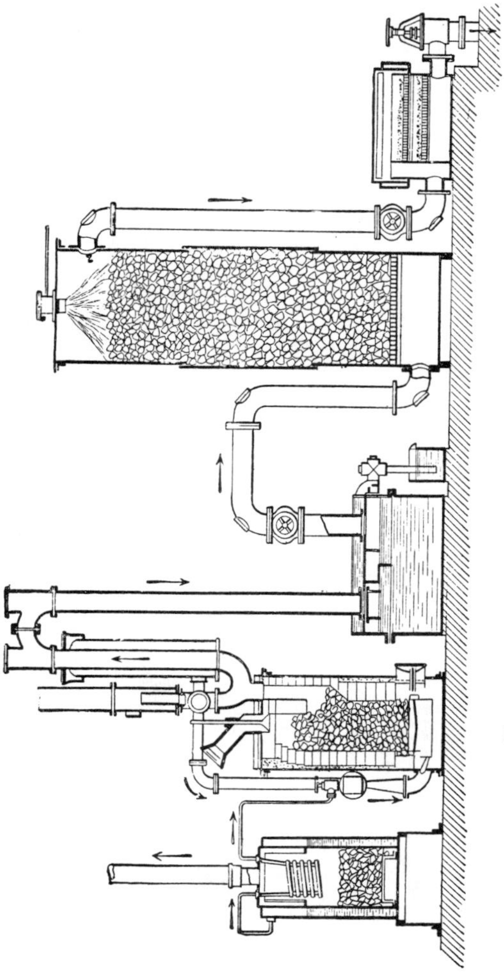

FIG. 77.—A complete Dowson producer-gas plant.

costly, and were, moreover, difficult to maintain in proper working order. Nevertheless, there are many cases in which they must be industrially employed.

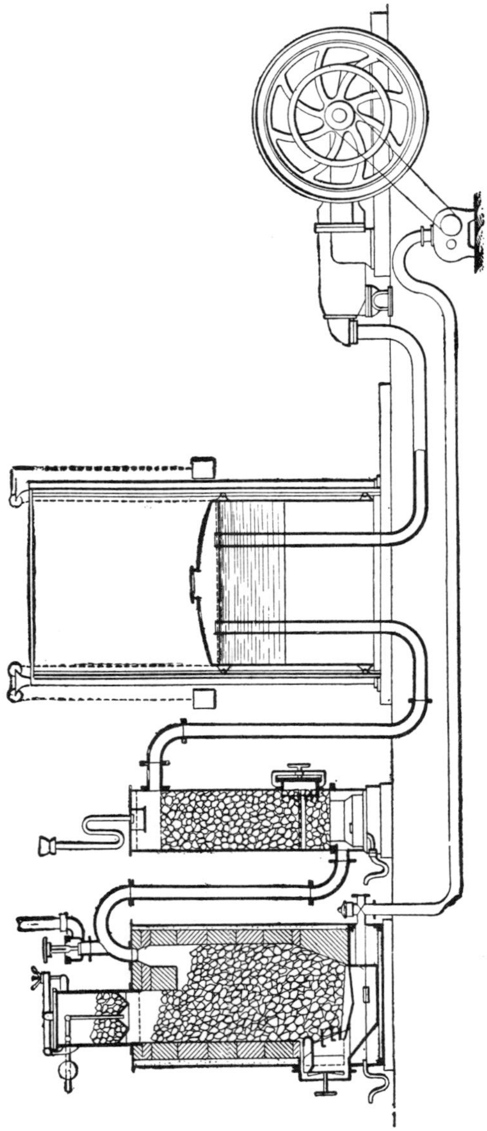

FIG. 78.—A Simplex producer-gas plant.

Among these may be cited works in which producer-gas is employed as a furnace fuel or as a soldering or roasting medium. Still other cases are those in which the producer-gas must be piped to some distance from a central generating installation to various engines, in the manner rendered familiar in gas-lighting practice.

Most pressure gas-generators have been copied from the original type invented by Dowson. These include a generator in which the gas is produced; an injector fed by a boiler; a fan or a compressor by means of which a mixture of steam and air is blown under the generator-furnace; washing apparatus termed "scrubbers"; gas-purifying apparatus; and a gas-holder (Fig. 77).

Generators.—The generator consists of a retort made of refractory clay, vertically mounted, and cylindrical or conical in form. This retort is protected on its exterior by a metal jacket with an intermediate layer of sand which serves to reduce the heat lost by radiation. The fuel is charged through the top of the retort, which is provided with a double closure in order to prevent the entrance of air during the charging operation. The generator rests on a grid arranged at the base of the retort, upon which grid the ashes fall. The outlet of the injector-pipe opens into the ash-pit, and this injector constantly supplies a mixture of steam and air. The mixture is generally superheated by passing it through a coil arranged in the fire-box of the boiler, in the generator, or in the outlet for burnt gases. Sometimes the air is subjected to a preliminary heating by recuperating in some way the waste heat of the apparatus.

The chief features in the arrangement of generators which have received the attention of manufacturers are the following: Good distribution of the fuel in charging; easy descent of the fuel; reduction of the destructive action of the clinkers on the walls; means for cleaning the grate without interfering with the generation of gas; prevention of leakage. Many devices have been employed to fulfil these requisites.

A perfect distribution of the fuel during charging is attained chiefly by the form of the hopper, and of its gate, which is generally conical. In most apparatus the gate opens toward the interior of the generator, and the inclination of its walls causes a uniform scattering of the fuel in the retort. It is all the more necessary to disperse the fuel in this manner when the cross-section of the retort is small compared with its height.

The facility of the fuel's descent is dependent largely upon the nature and the size of the coal employed. Porous coal gives better results than dense and compact coal. It is therefore preferable to employ screened coal free from dust in pieces each the size of a hazelnut. The various sections given to the interior, including as they do cylindrical forms, truncated at the summit or the base, partially truncated toward the base and the like, would lead to the conclusion that this question is not of the importance which some writers would have us believe. Still, it must be considered that if the fuel drops slowly, its prolonged detention within the walls of the hopper and its transformation into fusible slag may result in a disintegration of the refractory lining of the furnace.

The quantity of steam injected, greater or less, according to the nature of the fuel, renders it possible to obtain friable slags and consequently to prevent grave injury to the retort. Red-ash coal is in general fusible, containing as it does some iron. Its temperature of fusion varies between 1,832 to 2,732 degrees F.

Cleanliness is most important so far as the operation of the generator is concerned. It should be possible to scrape the generator during operation without changing the composition of the gas, when the incandescent zone is chilled, or an excess of air is introduced, or the steam-injector be momentarily thrown out of operation. Mechanical cleaners with movable grates or revolving beds have the merit of causing the ashes to drop without interfering with the operation of the apparatus. The same meritorious feature is characteristic of ash-pits having water-sealed joints.

Pressure gas-generators need not be as perfectly gas-tight as suction apparatus. Leakage of gas, which is usually manifested by a characteristic odor, results in a loss of consumption and renders the air unfit to breathe.

A generator should be provided in its upper part with openings through which a poker can easily be introduced in order to shake up the fuel and to dislodge the clinkers which tend to form and which cause the principal defects in operation, particularly with fuels that tend to swell, cake, and adhere to the furnace walls when heated. Many apparatus, moreover, are provided with lateral openings having mica panes through which the progress of combustion can be observed (Fig. 79).

Air-Blast.—The system by which air and steam are

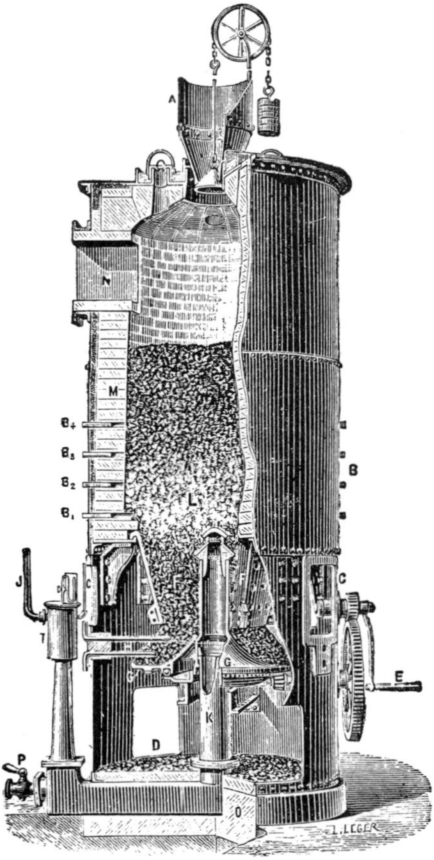

FIG. 79.—Fichet-Heurtey producer with rotating bed-plate.

injected necessitates the employment of a steam-boiler of 75 pounds pressure. This method of blowing, which is rather complicated, has the disadvantage of varying in feed with the pressure of the steam in the boiler, which pressure is not easily maintained at a given number of pounds per square inch. Moreover, when more or less resistance is offered by the fuel in the generator the quantity of air which is injected is likely to be diminished in quantity while the quantity of steam remains the same. The result is a change in speed which follows from the modification of the proportions of the two elements. For these reasons some manufacturers

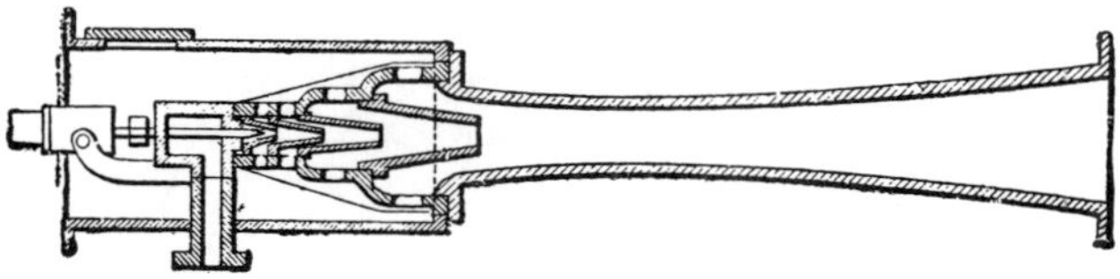

FIG. 80.—Koerting blower.

have resorted of late years to the employment of fans and blowers.

Blowers.—The fans or blowers employed vary considerably in arrangement. Most of them are based on the Koerting system (Fig. 80), and comprise essentially (1) a tube through which the steam is supplied under pressure, and (2) a cylindro-conical blast-pipe. The tube is placed in the axis of the blast-pipe at its outer opening. As it escapes under pressure the steam is caught in the blast-pipe and draws with it a certain quantity of air, which can be regulated. It is important that these injection blowers should operate in such a manner that the pressure and the feed of air and steam can be controlled.

Fans.—Mechanical blowers have the advantage of

dispensing with the employment of steam under pressure and the consequent installation of a boiler (Fig. 78). Driven by the engine itself or from some separate source of power, these apparatus are easily placed in position, require no great amount of attention, and utilize but little energy. They are either of the centrifugal type or of the rotary type, exemplified in the Root blower (Fig. 81). The latter system has the advantage

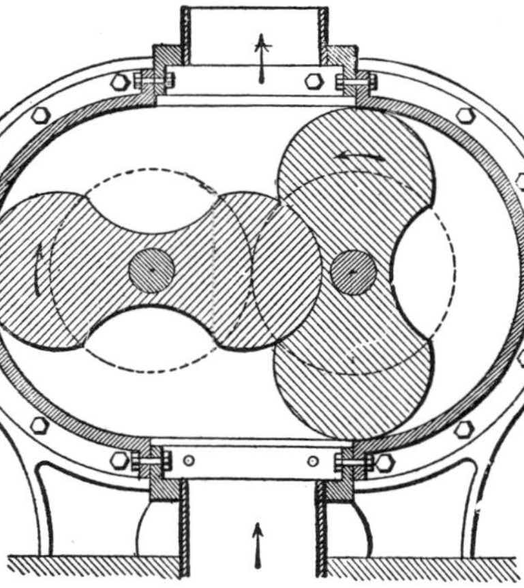

FIG. 81.—Root blower.

of high efficiency, and of enabling comparatively high pressures—19 to 27 inches of water—to be attained, which, however, are used only for special fuels, such as lignite, peat, and the like. The air supplied by the blower, before reaching the fire-box, is superheated, either before or after it is charged with steam.

Compressors.—In some installations air is supplied by compressor under the high pressure of 70 to 90 pounds per square inch, and seem well adapted to the production of a gas of good quality. Moreover, neither tar nor ammoniacal waters are produced. The Gardie producer may be considered typical of this class of apparatus (Fig. 82). The chief feature of this producer is to be found in simple washing and purifying apparatus. It may be well to state here that the compression of air at high pressure occasions some complications, and a considerable expenditure of power.

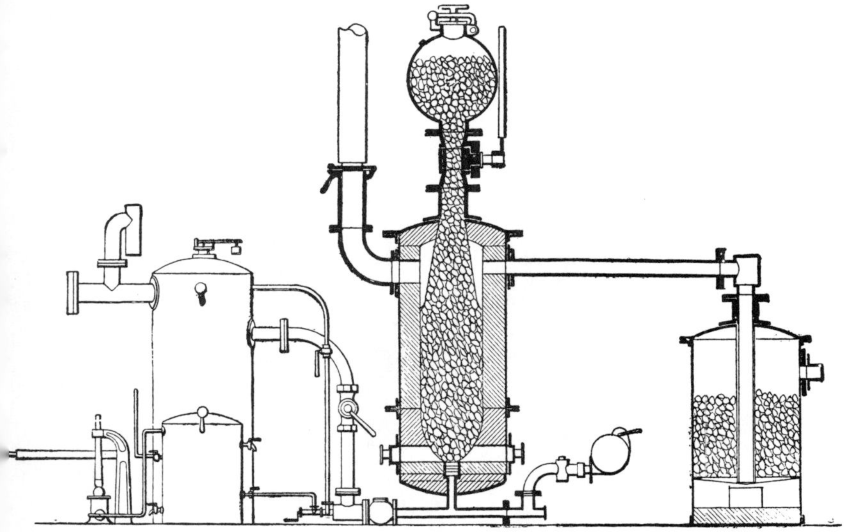

FIG. 82.—Gardie producer.

Exhausters.—Some designers have invented devices which draw gas into the generator whence it is supplied to the engines, these suction apparatus being connected with the blowers or used separately. But with the exception of a few special instances, such arrangements are not widely used—at least not for the production of motive power alone.

Whatever may be the arrangement employed for the

introduction of a mixture of air and steam under the grate of the generator, the blast-pipe as a general rule discharges toward the center of the apparatus. Still, in large producers it becomes desirable to provide a means for varying the quantity of air and steam within wide limits so as to regulate the heat of the fire. For that reason several outlets are symmetrically arranged below the fuel.

Washing and Purifying.—In pressure producers

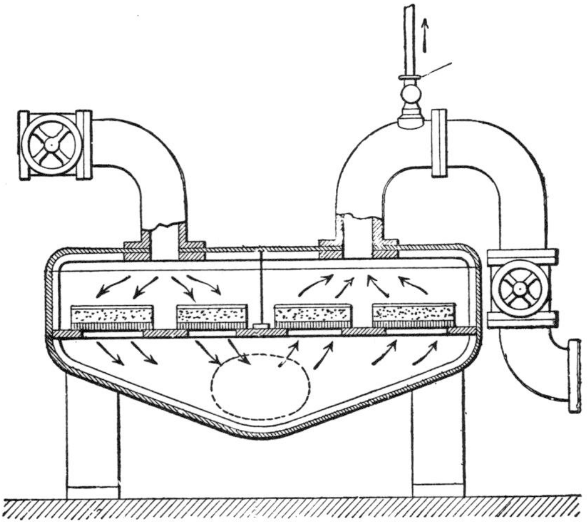

FIG. 83.—Sawdust purifier.

the gas is generally washed and purified with much more care than in suction apparatus. Given a sufficient pressure, the gas can be driven through the different apparatus and the spaces between the material which they contain without any difficulty. The gases emerge from the generator highly heated, and this heat is used either to warm the injection water or to generate the steam fed to the furnace. The gases then enter the washing apparatus, which most frequently consists of a succession of contrivances in which the gas is washed either by causing it to bubble up through the water, or by subjecting it to superficial friction against a sheet of water, or by systematically circulating it in a mass of continuously besprinkled inert material. The object of washing is to remove the dust contained in the gas and

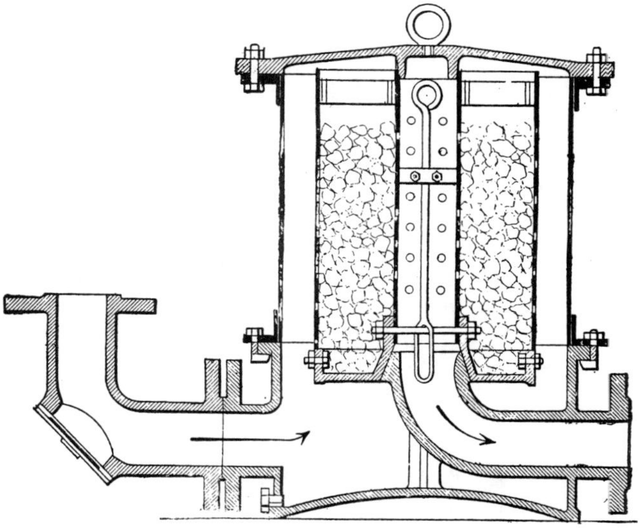

FIG. 84.—Moss or fiber purifier.

to precipitate it in the form of a slime which can be removed by flushing.

Physical purification thus begun is completed by passing the gas through a filtering bed consisting of fiber, sawdust, or moss (Figs. 83 and 84). Chemical purification, if it is necessary, is effected by means of calcium hydrate, iron oxide, or, still better, by a mixture of lime and iron sulphate. This filtering material must necessarily be renewed after it is exhausted.

Gas - Holder.—The gas-holder is composed essentially of a tank and a bell. Sometimes, for the purpose of simplifying the apparatus, the tank is so arranged as

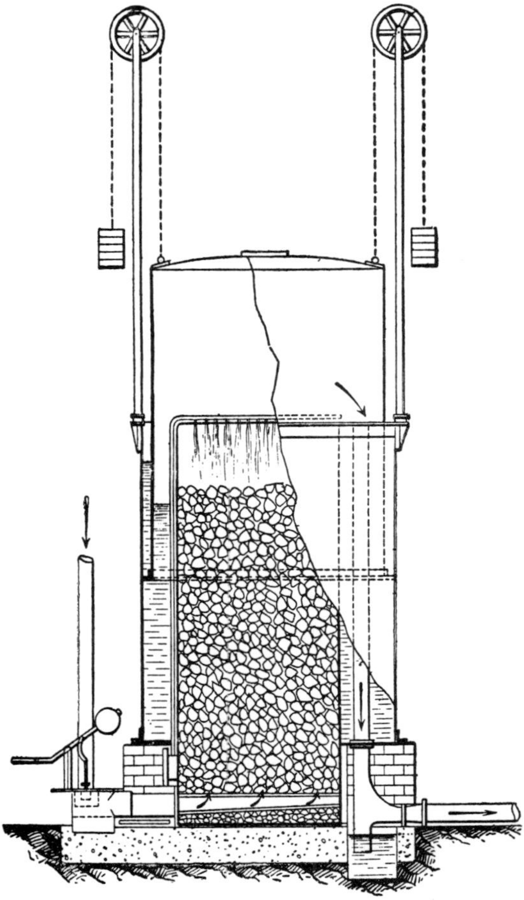

FIG. 85.—Combined gas-holder and washer.

to take the place of a washer or scrubber (Fig. 85). The bell should be provided with mechanism which, when the bell is full, automatically diminishes or stops the generation of gas. It is advisable to provide the bell with a blow or flap valve opening toward the interior. If, therefore, it should happen that the gas supply is cut off while the engine still continues to run, the suction of the engine will not draw the water from the tank of the gas-holder.

When engines are employed the horse-power of which does not exceed 50, it is sometimes customary to use the water of the tank (placed at a higher elevation than the engine) to cool the cylinder. In this manner the cost of installing special reservoirs is saved. If such an arrangement be employed, however, the quantity of water contained in the tank should be at least double that ordinarily contained in reservoirs. If this precaution be not observed, the water may become excessively heated and expand the gas in the bell.

The volume of the bell of the gas-holder should preferably be not less than about 3 cubic feet per effective horse-power of the engine to be supplied. Under these circumstances the bell acts as a pressure-regulator, assures a sufficient homogeneity of the remaining gas, and renders it possible to supply the engine during the short intervals in which it is necessary to stop the blast to poke the fire. But if the engine consumes 60 to 80 cubic feet of producer-gas per horse-power per hour, the bell must be very much larger in size if the generation of gas is to be checked for some time

It may be well to recall here that coal is not the only fuel which lends itself to the generation of gas suitable for driving engines, but that some generators are able to utilize lignite, peat, and the like. In others, straw, wood, shavings and sawdust, tannery waste, and other organic matter is burnt with an efficiency very much higher than that which they would give in the fire-boxes of steam-boilers.

Lignite and Peat Producers.—Lignite and peat generators (Fig. 86) cannot operate on the suction principle because of the resistance offered to the passage of gas by the layer of fuel. This resistance is considerable and extremely variable. Consequently, lig-

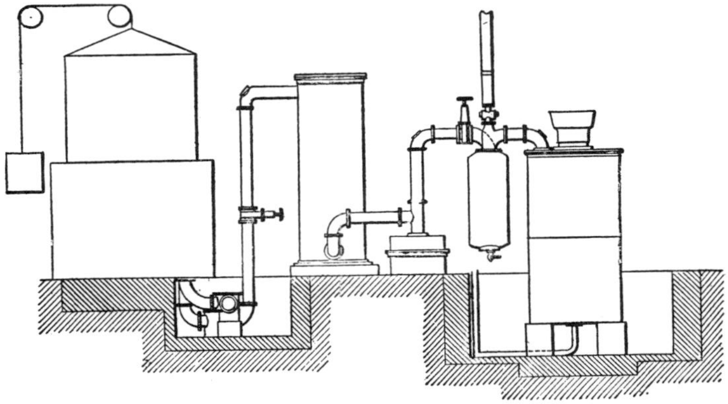

FIG. 86.—Otto Deutz lignite-producer.

nite and peat generators must operate on the pressure principle by utilizing a blast of air or a steam injector, depending upon the amount of water contained in the lignite. As a general rule a Root blower operating at a pressure of 8 to 27 inches of water, depending upon the quality of the lignite, is employed. These generators are not to be recommended for powers less than 50 horse-power, for the cost of the apparatus becomes too great.

The best lignite is that which, after combustion, leaves a fine ash and no agglomerated clinker. Lignite has the peculiarity of forming dust which ignites very easily when air is admitted into the generator. For this reason the generator should not be scraped during operation, in order to avoid the production of a flame which may escape from the apparatus.

The scrubber is simply a column without coke, and is provided with an interior sprinkler. The coke is too rapidly clogged with tar. Much of this tar is deposited in a chamber which precedes the gas-holder. Several quarts of tar may be tapped from the chamber daily.

The gas-holder serves merely to regulate the production of gas. The pipes leading to the engine should be cleaned several times each month, in order to remove the thin layer of tar which is deposited within them.

There are many kinds of lignite, and the gas-generator should be constructed to meet the peculiar requirements of the variety employed. The layer of fuel should be such in thickness that the gas as it emerges from the generator has a temperature of about 77 degrees F. This is the temperature of the gas which leaves the scrubber in the case of anthracite-generators. If the lignite contains much water, the greater part is retained in the washer by the gas in the form of drops. Sometimes the water drips through the grate of the generator. Lignite-generators may also be operated with peat, and even with town refuse, with slight modifications. The consumption per horse-power per hour is 3.3 pounds of lignite containing 2,400 calories (9,424.9 B. T. U.). In order to generate the same power with a boiler and steam-engine, 8.8 pounds would be required. An engine driven unloaded with fuel furnished by a lignite-generator will consume 50 per cent. of the weight of the fuel required at full load. This depends upon the proportion of water contained in the lignite and on losses of heat by radiation from the generator. In street-gas engines running without load, the absorption is 20 per cent., in anthracite-generators 40 per cent. of the consumption at full load.

Passing now to the utilization of wood, of which something has already been said in Chapter XI, two entirely distinct processes are successfully employed in apparatus of the Riché type, these processes depending upon the form of the wood used—whether, in other words, the wood be consumed in the form of sticks or blocks or in the form of chips, sawdust, bark, and the like, all of them the wastes of factories in which wood is used.

Distilling-Producers.—If the wood consists of logs, it is burnt in a generator comprising a fire-box and a distilling retort. The fire-box is charged with ordinary coal which serves to heat the retort to redness. The wood is discharged through the top of the retort, and the gas, produced by the distillation, escapes through the bottom and passes to the washing apparatus. The base of the retort is heated to about 1,652 degrees F., while at the top this temperature is reduced to 752 degrees F. The wood thus treated is transformed into charcoal, which is a by-product of some value.

The lower part of this cast retort (Fig. 87) is lined with charcoal, the residue of previous distillations.

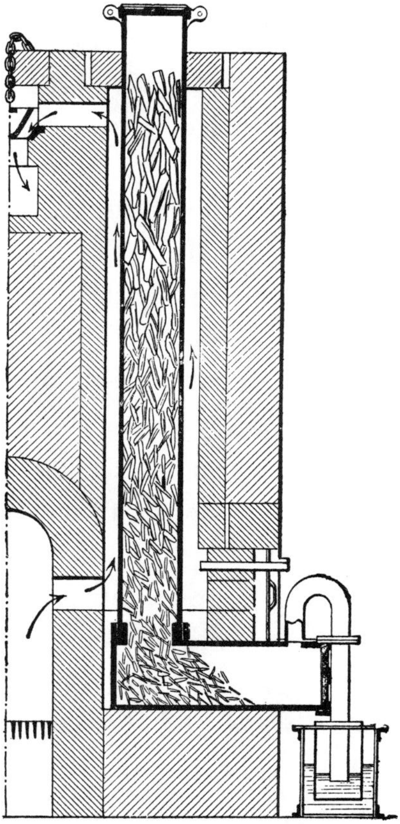

FIG. 87.—Riché distilling-producer.

The wood which is introduced in the upper part of the retort is distilled in the chamber. The retort is held by its own weight in a socket on the foot, which socket is lined with a special refractory cement, made of silicate, asbestos forming the joint. The products of combustion, issuing from the furnace, pass by way of the flue to the lower part of the casing, and raise the temperature of the retort and the charcoal it contains to that of a cherry red (1,652 degrees F.). These products of combustion then float to the upper part of the casing and heat the top of the retort to a temperature of about 752 degrees F., in which part the wood or the wooden waste to be distilled is enclosed. Thence the products of combustion pass through a horizontal flue, provided with a damper, into a collecting flue by which they are led to the smoke-stack. The products of distillation formed in the chamber, having no outlet at the top of the retort, must traverse the zone filled with incandescent carbon. The condensible products are conducted as permanent gases (carbonic-acid gas in the state of carbon monoxide) and are collected in the receptacle, after having passed the funnel and the bell of the purifying apparatus.

A gas-furnace is formed by grouping in a single mass of masonry a certain number of elements of the kind just described. It is essential that the retorts should be vertically placed, that they be made only of cast metal and not of refractory clay, and, finally, that their diameter be not much more than 10 inches, which size has been found most expedient in practice. The gas collected in the bell or in one or more of the receptacles passes into the gasometer and then into the service pipes. If 2.2 pounds of wood be distilled by burning in the furnace $\frac{8}{9}$ of a pound of coal of average quality or 2.2 pounds of wood (either sawdust or waste), 24.5 to 28 cubic feet of gas will be generated having a thermal value of 3,000 to 3,300 calories per cubic meter (11,904 to 13,094 B. T. U. per 35.31 cubic feet), and a residue .44 pounds of charcoal will be left.

In practice only the wood of commerce containing in the green state 20 to 40 per cent. of water, depending upon the variety, is used. Hornbeam contains the least water (18 per cent.), while elmwood and spruce contain the most (44 to 45 per cent.).

The blast apparatus of the generator being started, the gas is supplied under pressure. By reason of its permanent composition and its richness, it is an excellent substitute for street-gas in incandescent lighting, a good furnace fuel reducing agent.

Producers Using Wood Waste, Sawdust, and the Like.—If waste wood in the form of shavings, sawdust, straw, bark, and the like, should be employed, a still higher efficiency is obtained with self-reducing generators of the Riché type.

Combustion-Generators.—In combustion-generators (Fig. 88) the fuel is burnt and not distilled. The generator comprises two distinct elements. The first is the generator proper, in which the combustion takes place. Upon it is placed a hopper or fuel supply box. The second element is the reducer, in which by an independent process the reduction of the carbonic-acid gas, the dissociation of the steam, and the transformation of the hydrocarbons takes place. The generator is provided at its base with a grate having oblique bars in tiers, which grate is furnished with a channel in which the water for the generation of hydrogen flows. On a level with this grate, at the opposite side, is a flue communicating with the reduction column of coke. The incandescent zone of the generator should not extend above the level of the grate. Instead of passing through the layers of fresh fuel and out by way of the top, the gas generated flows directly into the reduction column where it heats the coke to incandescence. The high temperature to which the coke is subjected, coupled with the injection of air, effects useful reactions. This

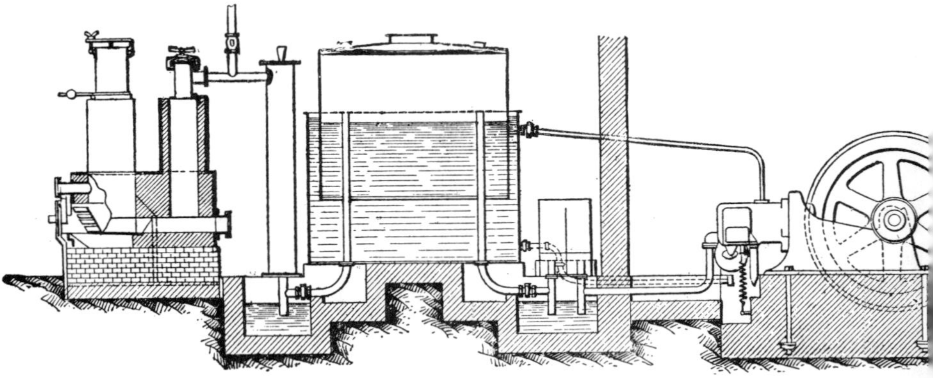

FIG. 88.—Riché combustion-producer.

additional air, however, is not used if the fuel is free from all products of distillation.

Experience has shown that gas of 1,000 to 1,100 calories per cubic meter (3,968 to 4,365 B. T. U. per 35.31 cubic feet), which heat content is necessary to develop one horse-power per hour, can be obtained with 3.96 pounds of wood in the form of shavings and sawdust containing 30 per cent. of water. The corresponding quantity of coke consumed in the reduction column is insignificant, and may be placed at about 0.112 pounds per horse-power per hour.

It has been proven in actual practice that, both in the distilling and combustion types of apparatus, the wood, either in the green state or in the form of saw-mill waste, may contain as much as 60 per cent. of water. Either of the two systems can be operated under pressure with an air-blast, in which case a gas-holder and bell must be employed. The gas as it passes from the generator to the gas-holder is conducted through a cooler and washer and through a moss filter, which removes traces of the products that may have escaped the distillation.

Inverted Combustion.—With a few exceptions the pressure-generators which have been described, as well as suction gas-producers which will be later discussed, are fed with anthracite coal or with coke. They cannot be operated with moderately soft or bituminous coal. For this reason they limit the employment of producer-gas engines. Manufacturers have long sought generators in which any fuel whatever can be consumed.

Among the producers which seem to overcome the objections cited to a certain degree, are those which are based on the principle of inverted combustion. These apparatus embody the ideas of Ebelmen, the products of distillation being decomposed by passing them over layers of incandescent fuel.

Many writers place in the class of inverted combustion producers, apparatus of the Riché, Thwaite, and Duff type, in which this idea is also carried out. Riché

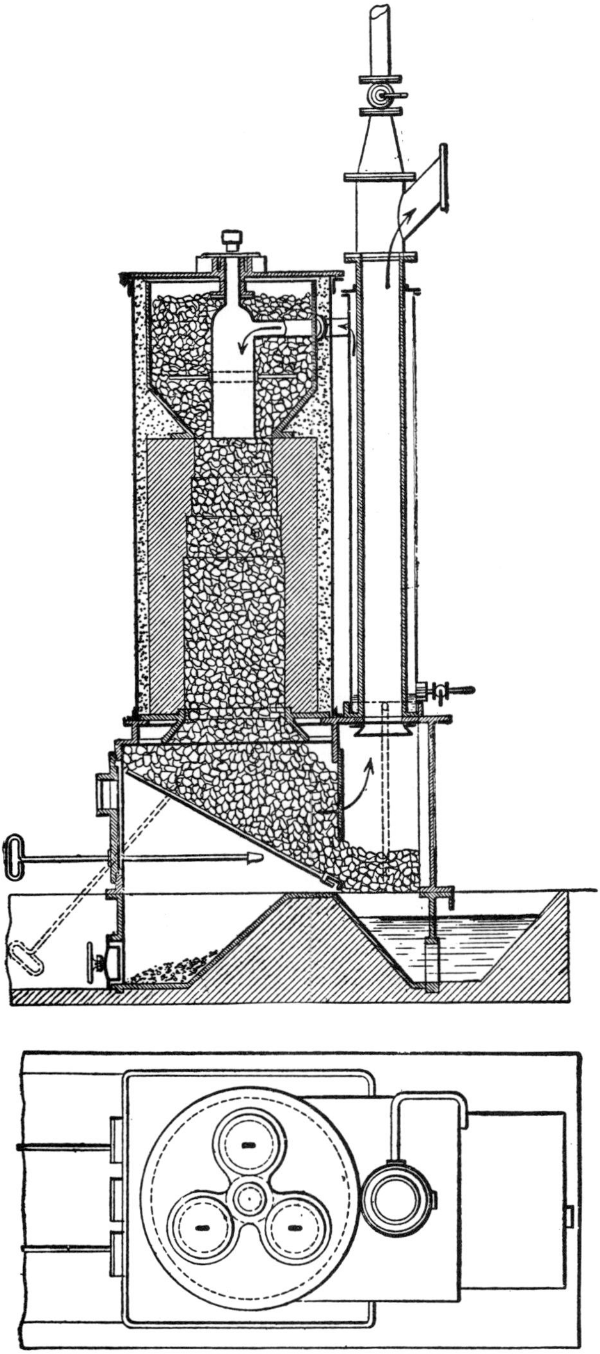

FIG. 89.—Deschamps inverted-combustion producer.

employs an independent incandescent mass to reduce the products of distillation of another mass. Thwaite employs two vessels which serve alternately as distilling

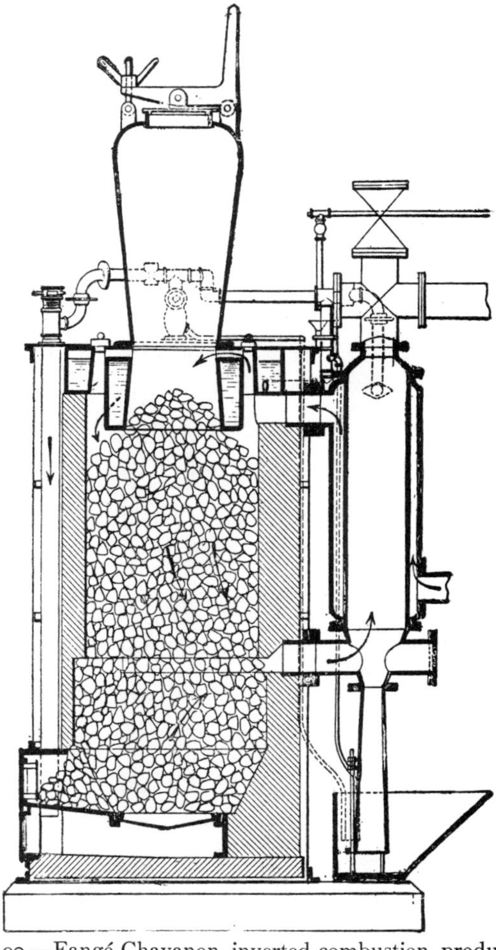

FIG. 90.—Fangé-Chavanon inverted-combustion producer.

retorts and reducing columns. Duff draws in the products of distillation for the purpose of blowing them under the fire. All these generators can hardly be said to be of the inverted combustion type.

The generators of Deschamps (Fig. 89) and of Fangé and Chavanon (Fig. 90), on the other hand, are producers in which the combustion is really inverted, and which are worked continuously. The air enters at the upper part of the retort, passes through the entire mass of fuel, carrying with it the distilled volatile products, and when the mixture reaches the incandescent zone, chemical reactions occur that result in the production of a gas entirely free from tar and other impurities.

CHAPTER XIII

SUCTION GAS-PRODUCERS

THE high cost and the complicated nature of the pressure gas-generators which have just been discussed have led manufacturers to attempt in some other way the generation of producer-gas intended for operating motors.

Several inventors, among whom we will mention Bénier and A. Taylor (in France), made some praiseworthy although not immediately very successful attempts to simplify the manufacture of producer-gas.

Advantages.—In these systems the suction occasioned by the motor itself has taken the place of a forced draft, produced in the generator by an air-injector or a fan, so that the gas, instead of being stored under pressure in a gas-holder, is kept in the apparatus under a pressure below that of the atmosphere.

As the device for producing a draft by means of boiler pressure or of a fan, and the gas-holder, are dispensed with, the result is a saving, first in the cost of installation, consumption, and floor space. Furthermore, the cooler and washer are supplanted by a single scrubber.

Manufacturers have succeeded in devising apparatus remarkable for the simplicity of the processes employed and yielding economical results which would never be obtained with pressure-generators employing gas-holders and boilers, considering that the boiler alone calls for a consumption of from 15 to 30 per cent. of the total amount of coal used for making the gas.

The best results obtained by the author with pressure gas-producers have indicated a consumption of not much less than 1 to 1¼ pounds of anthracite per horse-power per hour at the motor, while with suction-generators, under similar conditions and with the same grade of fuel, he has repeatedly found a consumption of from $\frac{9}{10}$ pounds per effective horse-power per hour. In either case, the gas obtained developed between 1,100 and 1,300 calories (4,365 and 5,158 B. T. U. per 35.31 cubic feet) if produced from anthracite yielding from 7,500 to 8,000 calories (29,763 to 31,746 B. T. U.) per 2.2 pounds.

The suction apparatus will also work very well with inferior coal containing up to 6 to 8 per cent. of volatile matter and from 8 to 10 per cent. of ash. This great advantage added to all the others explains the favorable reception which European manufacturers at once gave to suction-producers. The petroleum engine itself will find a serious competitor in the new system.

As regards the possibility of employing suction gas-generators with respect to the somewhat peculiar properties of the fuel, it may be said at the outset that coke from gas works yielding from 6,000 to 6,500 calories (22,911 to 24,995 B. T. U.) and also charcoal are perfectly available.

One horse-power per hour is obtained with a consumption of 1.1 to 1.3 pounds of coke.

Blast-furnace coke may be used in case of need, but its employment is not to be recommended on account of the sulphides it contains, which sulphides, being carried along by the gas, are liable to form sulphuric acid with the steam, the corrosive action of which would soon destroy the cylinder and other important parts of the engine.

Qualities of Fuel.—Anthracite coal is, upon the whole, so far the best available fuel for generators. However, it should possess certain qualities which will now be briefly indicated.

In suction gas-generators, above all, it is important that no harmful resistance should be opposed to the passage of the air and of the gas produced. It is therefore necessary to employ coal of a size that will answer the foregoing condition, without being too expensive.

The size of the pieces, to a certain extent, determines the price; and with coal of the same properties, pieces 1.1 to 2 inches may cost 1.4 of the price for the ordinary size of 0.59 to 0.98 inches, which is very well adapted for gas-generators. This is the size of a hazel-nut.

Moreover, it will be advisable to select the dryest coals, containing a minimum of volatile matter and having no tendency to coke or to cohere, in order that the volatilized products may not by distillation obstruct the interstices through which the gases must pass. For the same reason coal which breaks up and becomes pulverized under the action of the fire is not to be recommended. The coal should also be such as to avoid the formation of arches which would interfere with the proper settling of the fuel during its combustion. It may be stated as a rule that, with coal that does not cohere, the content of volatile matter should not exceed 5 to 8 per cent.

Coal which contains more than 10 to 15 per cent. of ash should not be used, for the reason that it chokes up and obstructs generators in which the dropping and discharge of the ashes is done automatically, a fact which should not pass unnoticed. The furnace cannot be cleaned safely with a fire of this kind, where combustion takes place in an enclosed space, without hindering the production of gas. Here again a point may be raised very much in favor of suction gas-producers. In a good generator, the ash-pit can be cleaned and the fire stoked without interrupting the liberation of the gas drawn in and without appreciably impairing the quality of the gas. These considerations are of importance so far as the gas-generator itself is concerned. Other conditions which should be noticed affect the engine fed by the generator, the grade of coal used, and the purification of the gas obtained from it.

Unless special chemical cleaners and purifiers are employed, thereby complicating the plant, the coal utilized should yield as little tar as possible during distillation; for the tendency of the tar to choke up the pipes and to clog the valves is one of the chief causes of defective operation of producer-gas engines.

Tar changes the proper composition of the explosive mixture. When it catches fire in the cylinder it causes premature ignition, which is so dangerous in large engines.

From what has been said in the foregoing, it follows that, in the present state of the art, the satisfactory operation of gas-generators depends no longer on the use of pure anthracite, such as Pennsylvania coal in America and Welsh coal in England, containing an amount of carbon as high as 90 to 94 per cent. and having a thermal value of 33,529 B. T. U. On the contrary, good dry coal yielding from 29,763 to 31,746 B. T. U. is quite suitable for the generation of producer-gas.

A final, practical advantage which speaks in favor of a generator and motor plant as compared with a steam-engine, is the small amount of water required. Apart from the water used for cooling the engine, which may be used over and over again if cooled, any water, whether it forms scale or deposits, may be employed for cooling and washing the gas in the scrubber.

According to the author's personal experience, an average of 3.3 gallons of water per effective horse-power per hour is sufficient for this purpose. This is about one-half of the amount required by a non-condensing slide-valve engine of from 15 to 30 horse-power. The difference in the consumption of water is quite important in city plants, where water is rather expensive as a rule.

General Arrangement.—A suction gas-generator

plant of the character we have been discussing is shown in Fig. 91.

The apparatus *A* is the generator proper, in which combustion takes place. The gas produced passes into the apparatus *B* through a series of tubes, to be conveyed to the washer *C*. In the apparatus *B*, which is the vaporizer, the water admitted at the top under atmospheric pressure is vaporized by contact with a series of tubes, heated by the gas coming from the generator.

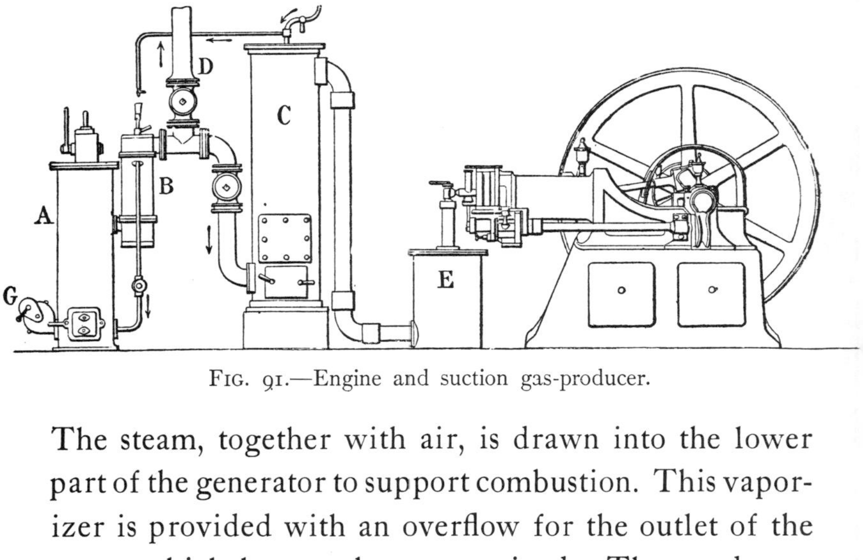

FIG. 91.—Engine and suction gas-producer.

The steam, together with air, is drawn into the lower part of the generator to support combustion. This vaporizer is provided with an overflow for the outlet of the water which has not been vaporized. The producer-gas pipe which leads from the vaporizer to the washer has a branch *D*, for the temporary escape to the atmosphere of the gas produced before and after the operation of the engine. In the washer, as the drawing shows, the gas enters at the bottom and leaves at the top to pass to the gas expansion-chamber *E* and thence to the motor. The gas thus passes through the body of coke in the opposite direction to the wash water, which then flows to the waste-pipe. The coke and the water free the gas not only from the dust carried along, but from the ammonia and other impurities contained in the gas.

When firing the generator, a small hand ventilator *G* is used for blowing in air to fan the fire. The gas obtained at first, being unsuitable for combustion, is allowed to escape through the branch *D*. After injecting air for about 10 to 15 minutes, the engine can be started after closing the branch *D*. The suction of the engine itself will then gradually bring about the proper conditions for its regular running, and after a quarter of an hour or half an hour the gas is rich enough to run the engine under a full load.

The apparatus just described is the original type, upon which many improvements have been made for the purpose of securing a uniform gas production and of diminishing the interval of time elapsing between the firing of the generator and the running of the engine under a full load.

Each of the elements of this apparatus—to wit, the generator, vaporizer, superheater, and washer—have been modified and improved more or less successfully by the manufacturers; and in order that the reader may perceive the merits and the drawbacks of the various arrangements adopted, the most important ones will be separately discussed.

Generator.—With respect to the general arrangement of parts, generators may be divided into two classes:

First.—Generators with internal vaporizers, such as the Otto Deutz and Wiedenfeld generators.

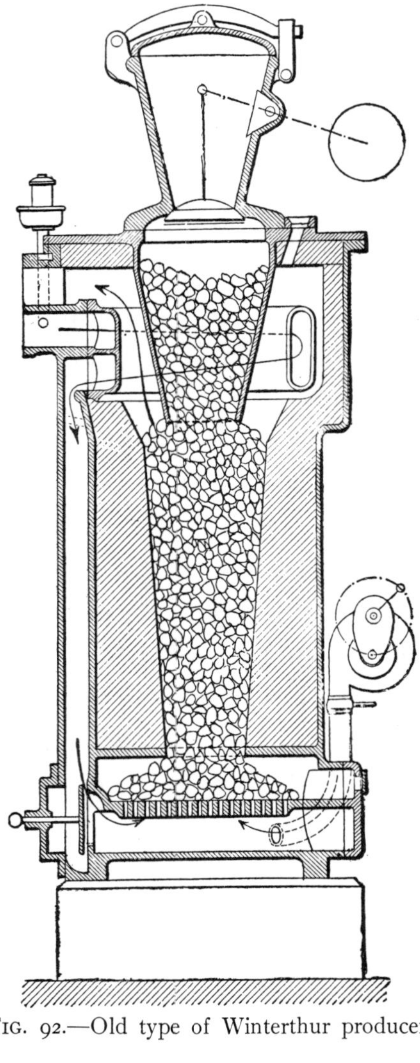

FIG. 92.—Old type of Winterthur producer.

Second.—Generators with external vaporizers, such as the Taylor, Bollinckx, Pintsch, Kinderlen, Benz, Wiedenfeld, Hille, and Goebels generators.

Cylindrical Body.—The generator consists essentially of a mantle made of sheet-iron or cast-iron and containing a refractory lining which forms a retort, a grate, and an ash-pit. In the small size apparatus the cast-iron mantle is often used, whereas in large sizes the mantle is made of riveted sheet-iron so as to reduce its weight and its cost. In the latter case the linings are securely riveted or bolted.

The Winterthur generator (Figs. 92 and 93), the Taylor generator (Fig. 94), and the Benz generator (Fig. 97), are made of cast-iron; the Wiedenfeld generator (Fig. 95), the Pintsch generator (Fig. 96), are made of sheet-iron; the Bollinckx (Fig. 98) is made partly of sheet-iron and partly of cast-iron.

The different parts of a generator, if made of sheet-iron, are held together by means of angle-irons forming yokes, and a sheet of asbestos is interposed. If the parts are made of cast-iron, they are connected after the manner of pipe-joints and packed with compressed asbestos. This latter way of assembling the parts presents the advantage of allowing them to be dismembered readily. Therefore, it allows the several parts to expand freely and facilitates the securing of tight joints. This last consideration is exceedingly important, particularly for the joints which are beyond the zone in which the distillation of the fuel takes place. Any entrance of air through these joints would necessarily impair the quality of the gas, either by mingling therewith, or by combustion. The air so admitted would also be liable to form an explosive mixture which might

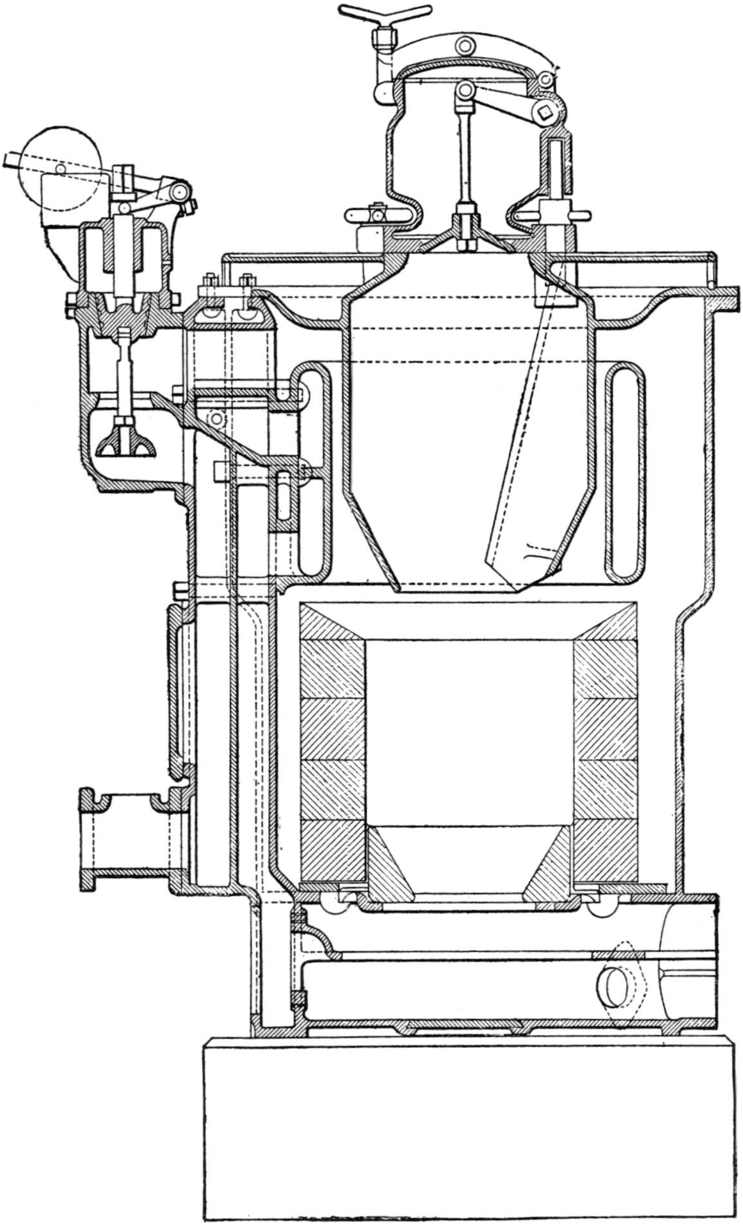

FIG. 93.—New type of Winterthur producer.

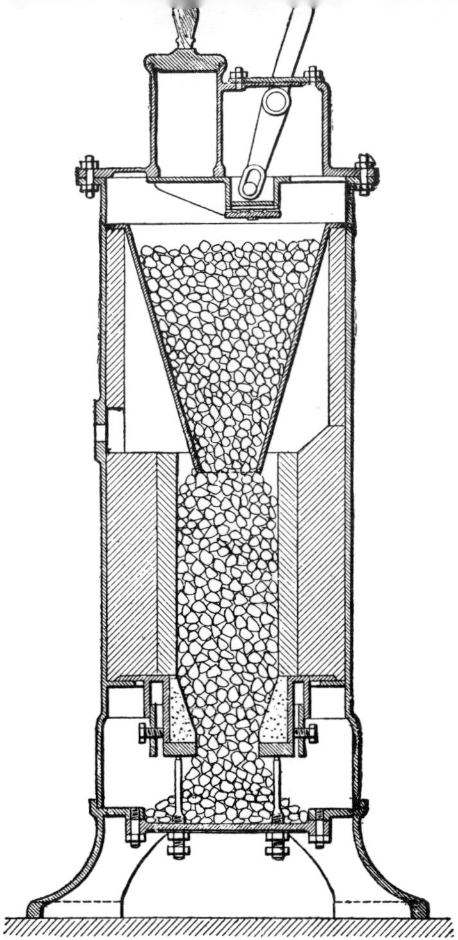

FIG. 94.—The A. Taylor producer.

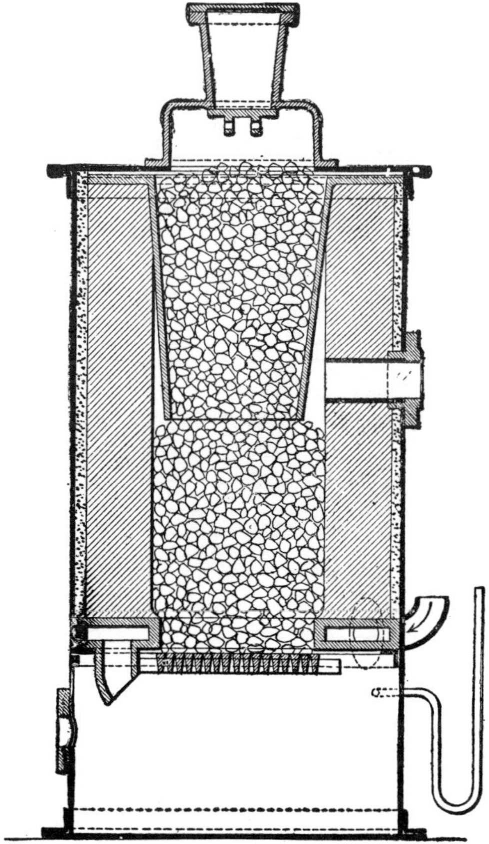

FIG. 95.—Wiedenfeld producer.

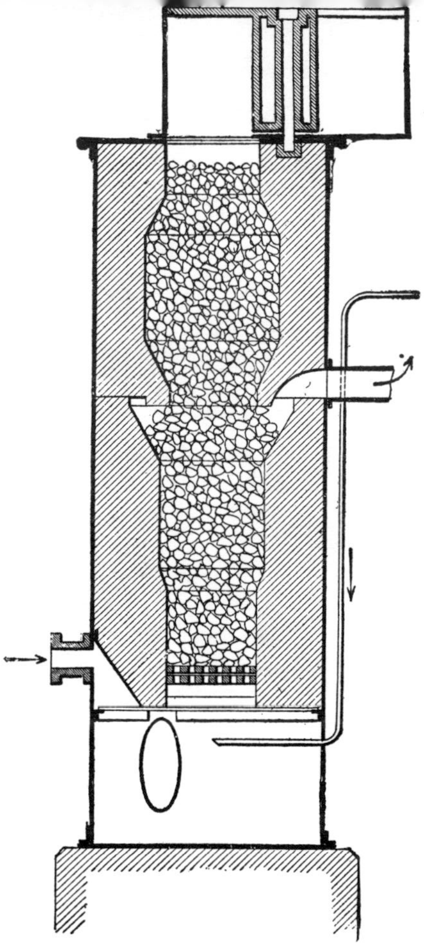

FIG. 96.—Pintsch producer.

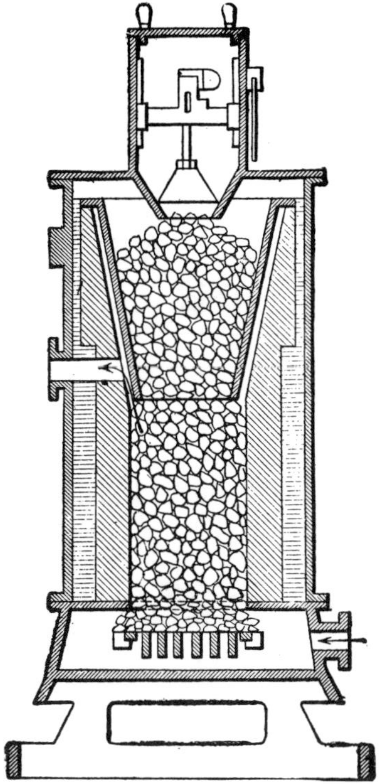

FIG. 97.—Benz producer.

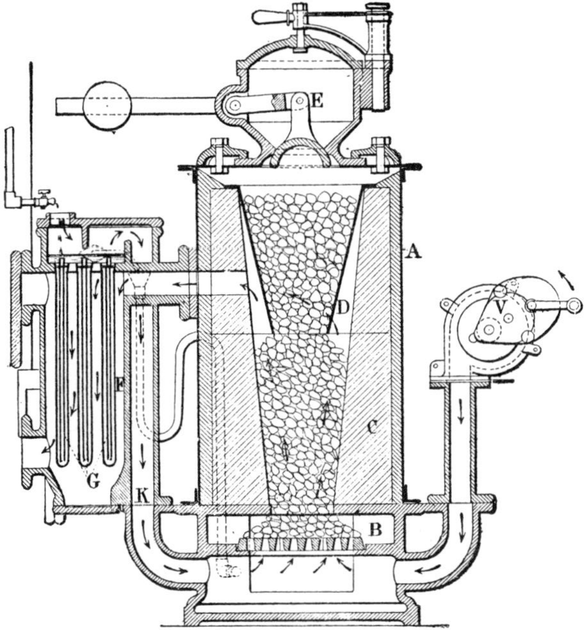

FIG. 98.—Bollinckx producer.

become ignited in case of a premature ignition of the cylinder charge during suction or through some other cause.

Refractory Lining.—The interior lining of the generator should be made of refractory clay of the best

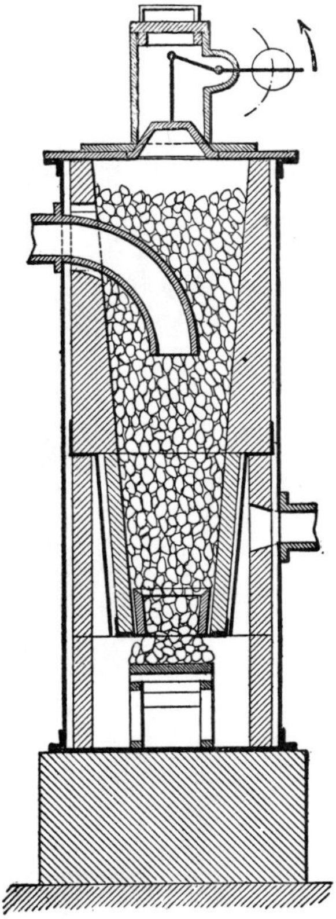

FIG. 99.—Lencauchez producer.

quality. It would seem advisable, in order to facilitate repairs, to employ retorts made of pieces held together instead of retorts made of a single piece. In the first case the assembling should preferably be made by means of refractory cement, and the inner surface should be covered with a coating so as to form a practically continuous stone surface.

Some manufacturers, in order to allow for the re-

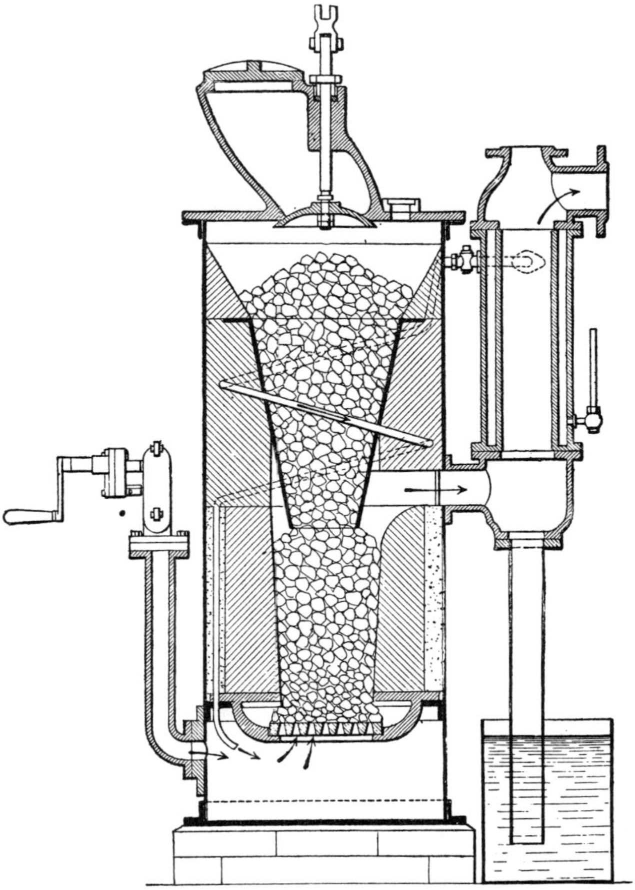

FIG. 100.—Goebels producer.

newal of the part most liable to be burnt, employ at the bottom of the tank a refractory moulded ring (Lencauchez, Fig. 99).

It is always advisable to place between the shell or

mantle of the generator and the refractory lining a layer of a material which is a bad conductor of heat as,

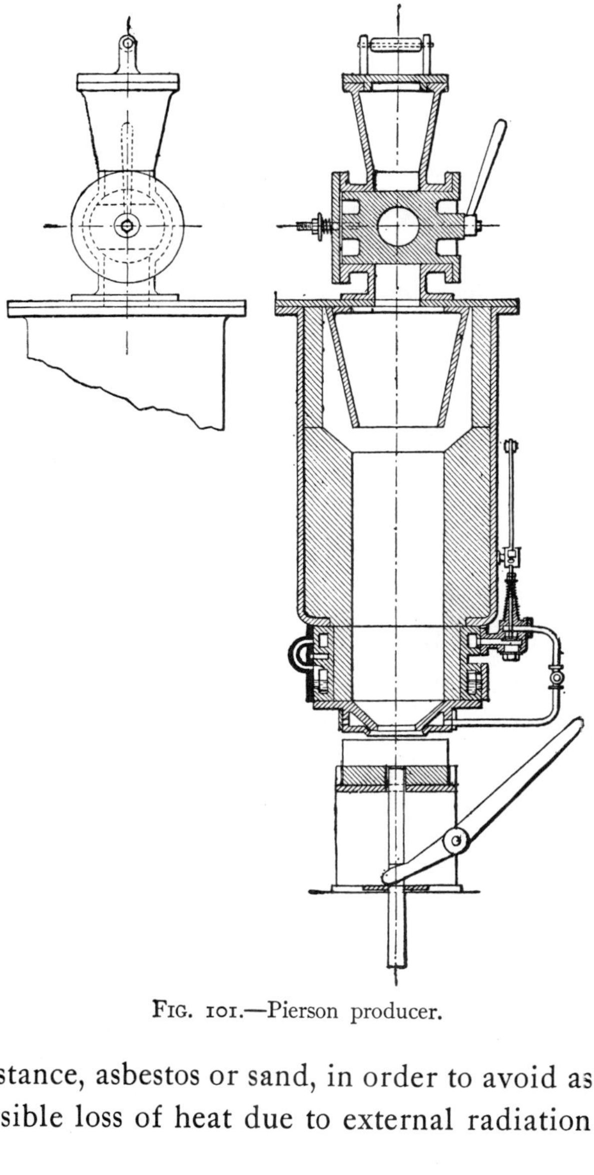

FIG. 101.—Pierson producer.

for instance, asbestos or sand, in order to avoid as much as possible loss of heat due to external radiation (Fig. 100).

Grate and Support for the Lining.—These parts, owing to their contact with the ashes and the hot embers, are liable to deteriorate rapidly. It is therefore indispensable that they should be removable and easily accessible, so that they may be renewed in case of need. From this point of view, grates composed of independent bars would appear to be preferable. The clearance between the bars depends, of course, on the kind of ashes resulting from the different grades of fuel. It is advisable to design the grate so that the free passage for the air is about 60 to 70 per cent. of the total surface.

In generators having a cup-shaped ash-pit, containing water (Fig. 95), the grate and the base of the retort are less liable to burn than in apparatus having dry ash-pits. Certain apparatus, such as those of Lencauchez (Fig. 99), Pierson (Fig. 101), and Taylor (Fig. 94), have no grates; the fuel is held in the retort by the ashes, which form a cone resting on a sheet-iron base, easy of access for cleaning and from which the fuel slides down gradually.

The Pierson generator (Fig. 101) is provided with a poker comprising a central fork, which is worked with a lever, in order to stir the fire from below without entirely extinguishing the cone of ashes.

In some apparatus in which a grate is used (Fig. 92), a space is left between the grate and the support of the retort. This arrangement has the merit of allowing only finely divided and completely burnt ashes to pass to the ash-pit. Moreover, a large surface grate can be employed, thus facilitating the passage of the mixture of air and steam.

The space above mentioned is provided with a clean-

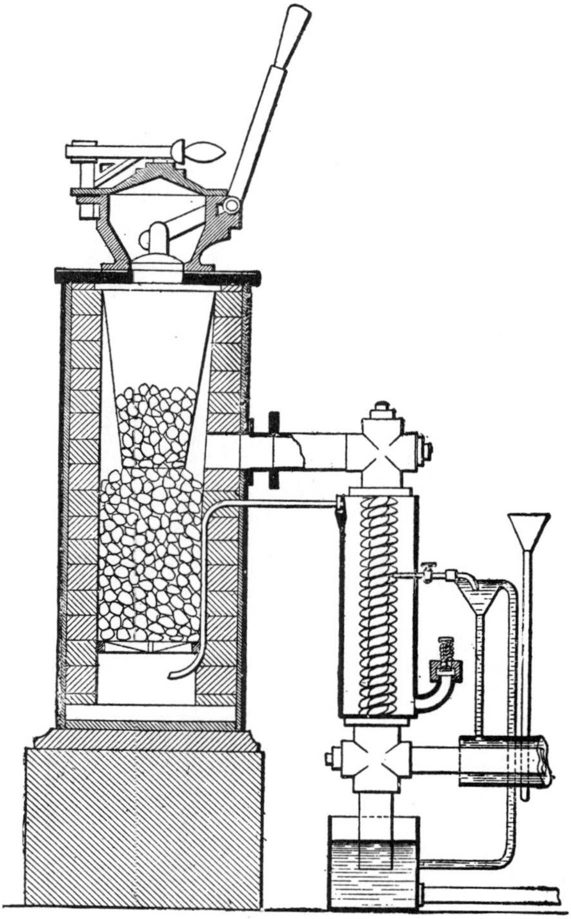

FIG. 102.—Kiderlen producer.

ing-door through which cinder and slag may be removed.

In other apparatus the grate rests either on the support of the refractory lining, as in the old type invented by Wiedenfeld (Fig. 95), or upon a projection embedded in the lining, as, for instance, in the Kiderlen (Fig. 102) and Pintsch generators (Fig. 96).

In the Riché apparatus (Fig. 103) there is, besides the ordinary grate, a grate with tiers on which the fuel

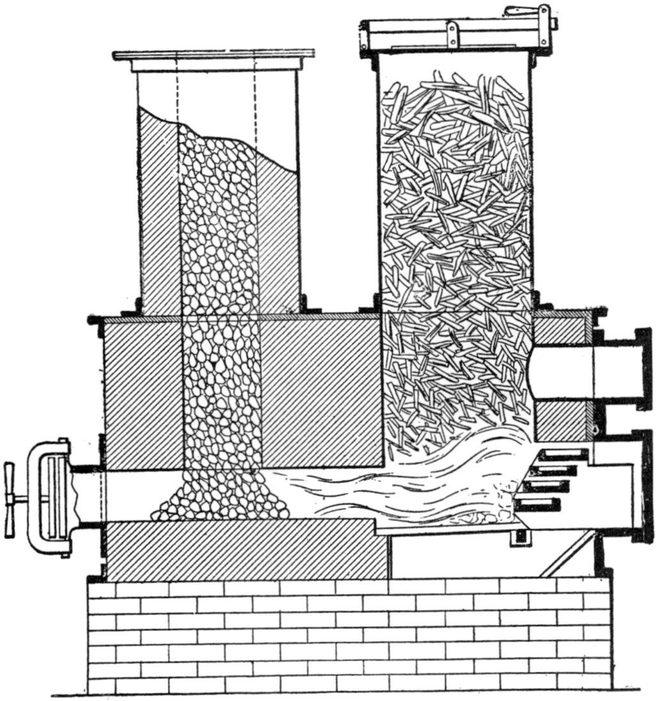

FIG. 103.—Riché combustion-producer.

spreads. This grate consists of wide, hollow bars containing water. It should be noted that the apparatus is of the blower type.

An interesting arrangement is found in Bénier's generator (Fig. 104). This consists of a grate formed of projections cast around a cylinder which can be turned about its axis. The finely divided ashes which are retained in the spaces between these projections are thus carried into the ash-pit, and those which adhere to the metal are scraped away by a metallic comb fastened to

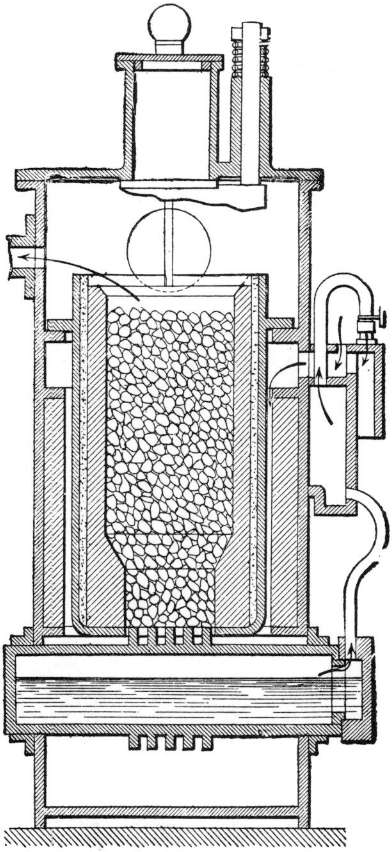

FIG. 104.—Bénier producer.

the lower part of the apparatus. The "Phœnix" generator (Fig. 105) is fitted with a grate having a mechanical cleaning device, worked by a lever from the outside.

Ash-Pit.—The ash-pits are exposed to the destructive

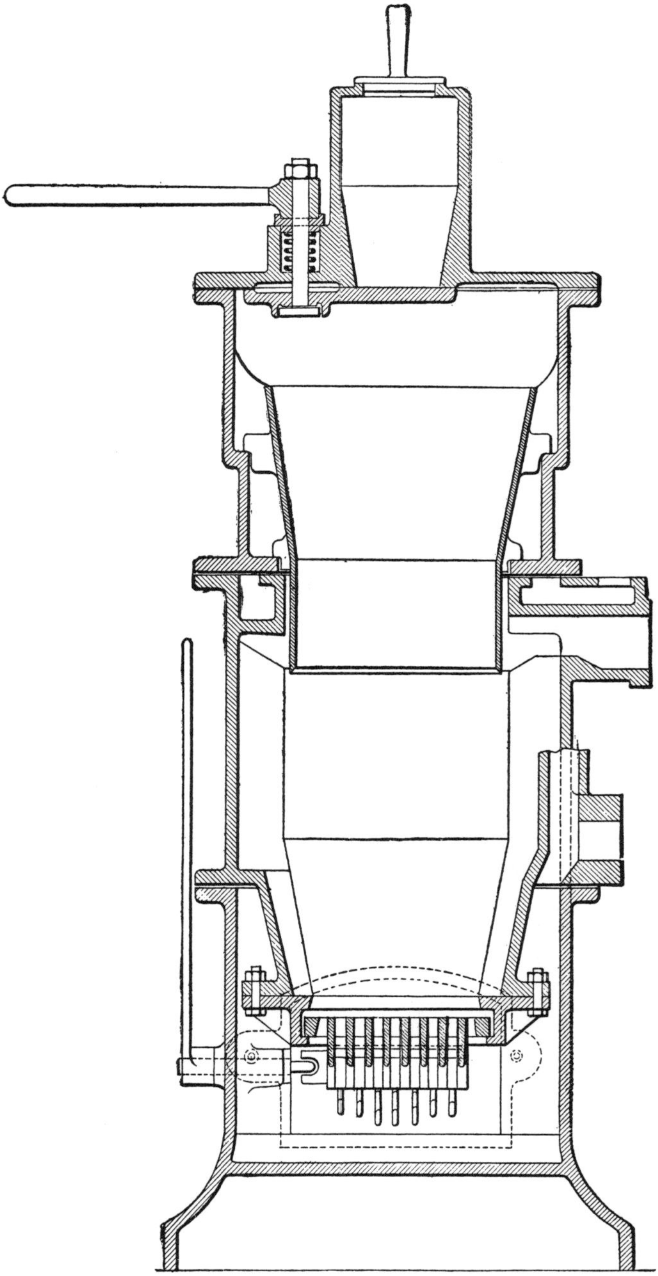

FIG. 105.—Phœnix producer.

effects of heat and moisture, and should preferably be constructed of cast-iron, since sheet-steel is liable to corrode quickly.

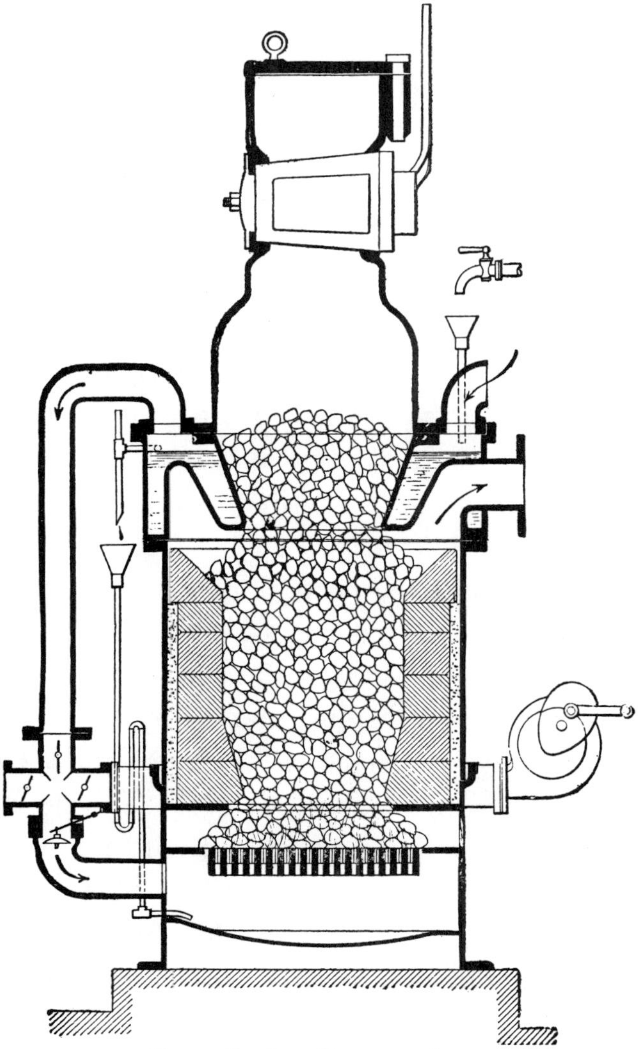

FIG. 106.—Otto Deutz producer.

In most apparatus the ash-pit is hermetically sealed, and the air for supporting combustion enters below the grate through a pipe leading from the heater or the vaporizer. This arrangement seems best adapted to prevent the leakage of gas which tends to take place by reaction after each suction stroke of the engine.

Ash-pits formed as water-cups, such as the Deutz (Fig. 106), the Wiedenfeld (Fig. 95), and the Bollinckx (Fig. 98), are fed by the overflow from the vaporizer. These ash-pits are themselves provided with an overflow consisting of a siphon-tube forming a water-seal.

Besides providing protection to the grate and other parts by this sheet of water, a larger proportion of the heat radiated from the furnace is utilized for the production of steam which contributes to enrich the gas. The doors of the ash-pits and their fittings are likewise exposed to a rapid deterioration.

For this reason these parts should be very strongly made, either of cast-iron or cast-steel. Furthermore, they should, at joint surfaces, be connected in an air-tight manner, which may be attained by carefully finishing the engaging surfaces of the frame and the door proper, or by cutting a dovetail groove in one of the sides of the frame which is packed with asbestos and adapted to receive a sharp edged rib on the other part.

The pintles of the hinges should also be carefully adjusted so that the joint members of the door shall remain true. Hinges with horizontal axes seem to be preferable in this respect to those having vertical axes. As a means of closing the door, the arrangement here shown (Fig. 107) seems to assure a proper engagement of the joint surfaces. It consists of a yoke which straddles the door, and which, on the one hand, swings about the hinge, and on the other hand engages a movable hoop. A screw, fastened to the yoke, serves to tighten the door by pressure on its center. This screw can also be fastened to the end of the yoke (Fig. 108).

It is very advantageous to provide in each door a hole closed by an air-tight plug, so that in case of

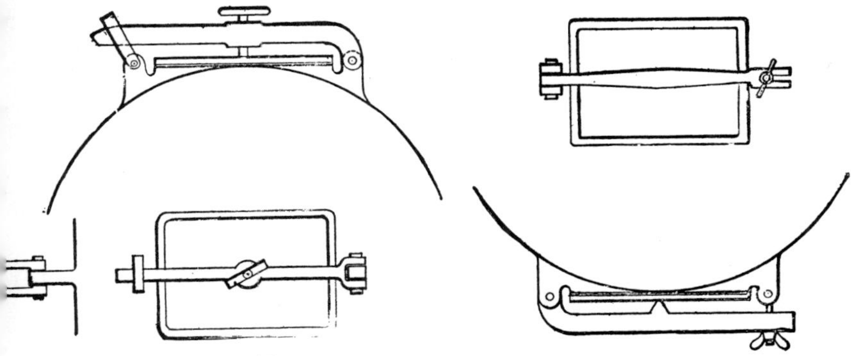

FIGS. 107–108.—Fire-box doors.

need a tool may be introduced for cleaning the grate. In this manner the grate may be cleaned without opening doors and without causing a harmful entrance of air.

The door of the furnace, particularly, should be provided with an iron counter-plate held by hinged bolts (Fig. 109); or, better still, this door should be so constructed that it can be lined with refractory material to protect it against the radiated heat of the fire.

Charging-Box.—Like the other parts of the generator the construction of which has been discussed above, the charging-box should be absolutely air-tight.

On account of their greater security, preference should be given to double closure devices, which form a sort of preliminary chamber, owing to which the filling of the generator is made in two operations. The first operation consists in filling the preliminary chamber after opening the outer door. Upon closing this outer door, the second operation is performed, which consists in moving the inner door so as to cause the fuel in the preliminary chamber to drop into the generator.

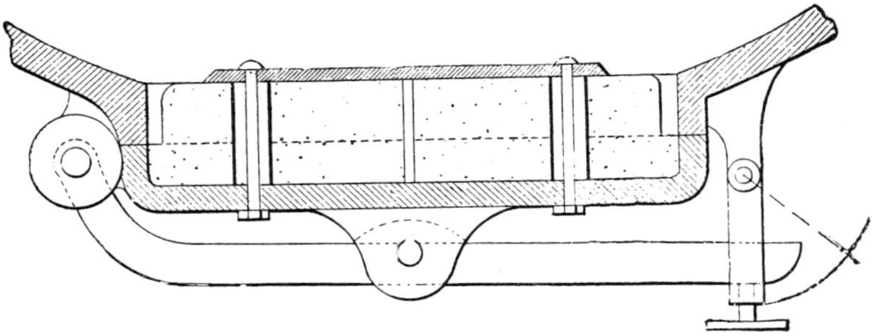

FIG. 109.—Door with refractory lining.

Stress has been laid on the greater safety of this type of charging-box for the reason that, with devices having a single charging-door, a sudden gust of air may rush in at the time of charging the furnace, and bring about an explosion very dangerous to the workman entrusted with stoking the furnace.

The closure is generally simply a removable cover, or may be a lid swinging about a hinge having a horizontal or vertical axis.

As regards the inner door, which is of great importance, in order to insure an air-tight joint, there are three chief types of closure:

1. The Lift-Valve.
2. The Slide-Valve.
3. The Cock.

The Lift-Valve.—The lift-valve is formed by a disk of conical or spherical shape moved up and down by means of a lever having a counter-weight for adjustment. The valve is used in the Winterthur (Fig. 92) and Bollinckx (Fig. 98) generators.

This device serves as an automatic closure and insures a tight joint irrespective of wear. Moreover, it presents the advantage that, at the moment of opening, it distributes the fuel evenly in the generator; but on the other hand, it has the drawback of not allowing the fuel to be examined or shaken through the charging-box. In apparatus provided with this kind of valve, it is therefore advisable to furnish the upper part of the generator with agitating holes closed by an air-tight slide.

Slide-Valve.—The slide-valve closure consists of a smooth-finished metallic plate movable below the charging-box proper. Operated as it is from the outside, it is evident that the slightest play, the wearing of the pivot, or the weight of the charge, will form spaces between the plate and its seat through which air may rush in.

Furthermore, the manipulation of the slide-valve may be interfered with if too much fuel is put in the generator.

The valve or damper may move parallel to itself or swing about the operating axis. The Taylor apparatus

(Fig. 94) and the Bénier apparatus (Fig. 104) are provided with such valves.

The Pintsch generator (Fig. 96) is provided with a device which, properly speaking, is not a damper, but which consists of two boxes movable about a vertical axis and arranged to be displaced alternately above the shaft to effect the charging. This system effects only a single closure, but explosions are scarcely to be feared with an apparatus of this kind, owing to the considerable height of fuel contained between the charging opening and the gas-producing zone.

Cock.—The cock is applied particularly in the modern apparatus of the Otto Deutz Co. (Fig. 106) and the Pierson generator (Fig. 101). It consists of a large cast-iron cone, having an operating handle and an opening. The cone moves in a sleeve formed by the charging-box.

This arrangement appears to be preferable to the others on account of its simplicity and of the ease with which it can be taken apart for cleaning. Moreover, the fuel can be poked directly through the feed-hopper. In apparatus provided with a cock, it is advisable to place on the outside cover a mica pane through which the condition of the fuel may be examined without danger.

Feed-Hopper.—Below the charging-box is arranged, as a rule, a hopper tapered conically downward. This part of the generator should serve only as a storage chamber for fuel. It can therefore be made of cast-iron, and has the advantage of being removable, easily replaced, and of allowing ready access to the retort for the purposes of examination and repair.

The annular space surrounding this feed-hopper generally forms a chamber for receiving the gas produced, as in the Winterthur (Fig. 92), the Bollinckx (Fig. 98), and the Taylor apparatus (Fig. 99).

In generators having an internal vaporizing-tank, this tank itself serves as a feed-hopper, which is the case in the Deutz apparatus (Fig. 106) and Wiedenfeld generator (Fig. 95).

Connection of Parts.—In order to facilitate the thorough cleaning of the retort, preference is given to removable charging-boxes and feed-hoppers. These are features of apparatus of the Bollinckx type (Fig. 98), in which the charging-box is secured to the generator by means of its yoke and by catches provided with knobs, and also of apparatus of the Winterthur kind (Fig. 92), having a charging-box pivoted about a vertical axis, or apparatus of the Duplex type (Fig. 110), in which the charging-box can swing about a horizontal hinge.

Air Supply.—We have seen that, when starting the generator, the gas is produced with the aid of a fan. This fan may be operated mechanically, but is generally operated by hand.

It is customary to convey the air-blast through a pipe leading to the ash-pit, as in the Winterthur apparatus (Fig. 92). Often, however, the air supply pipe is directly branched on that which leads from the vaporizer to the ash-pit, as in the Deutz apparatus (Fig.

106). In this case a set of valves or dampers permits the disconnection of the fan or its connection with the ash-pit.

In some apparatus an air inlet is provided immediately adjacent to the ash-pit. This arrangement is faulty for the reason that it gives rise to gaseous emanations which take place by reaction after each suction

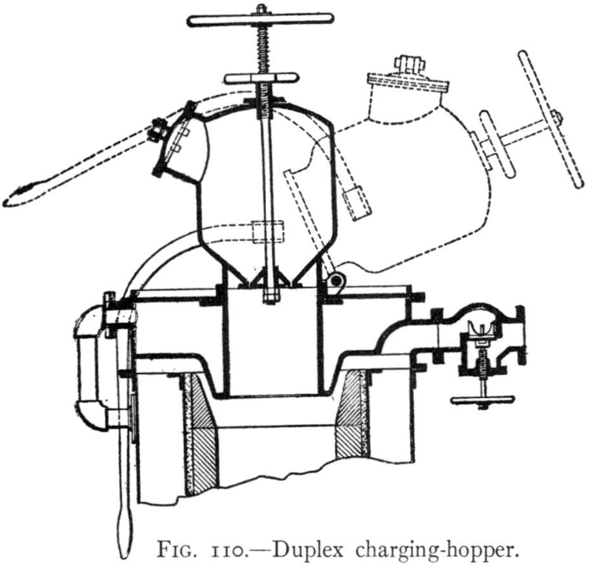

FIG. 110.—Duplex charging-hopper.

stroke of the engine. Furthermore, it is advisable that the air supplied below the ash-pit be as hot as possible. For this reason the employment of preheaters is desirable. The dry air forced in by the fan stimulates combustion, and the hot gas produced and mixed with smoke escapes through a separate flue, generally arranged beyond the vaporizer and serving as a chimney. This chimney should in all cases be extended to the outside of the building, and should never terminate in a brick chimney or similar smoke-flue. The direct escape of such gas and smoke through a telescopic

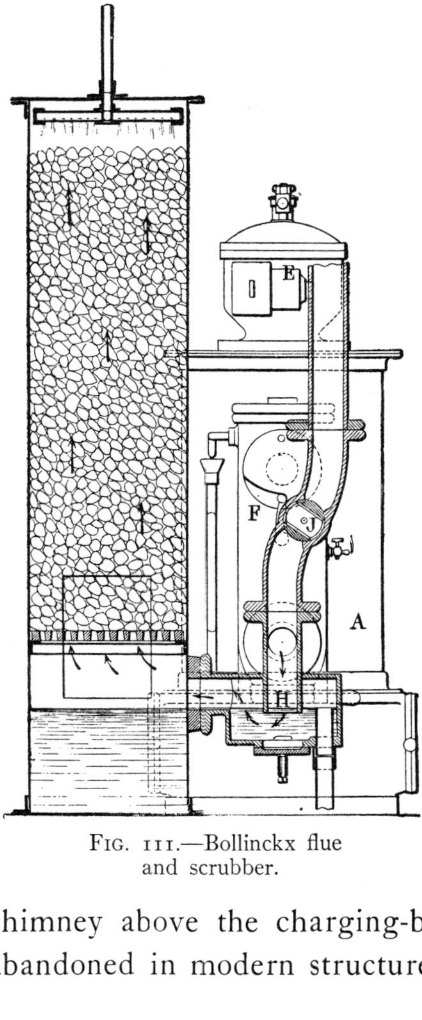

FIG. 111.—Bollinckx flue and scrubber.

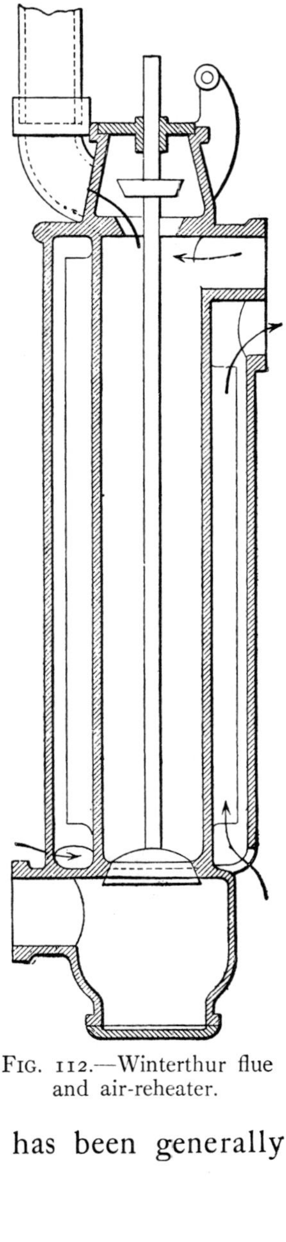

FIG. 112.—Winterthur flue and air-reheater.

chimney above the charging-box has been generally abandoned in modern structures.

The escape-pipe mentioned, being branched on the gas-pipe leading to the engine, should be capable of disconnection when desired, by a thoroughly tight system of closure. For this purpose, some employ a simple

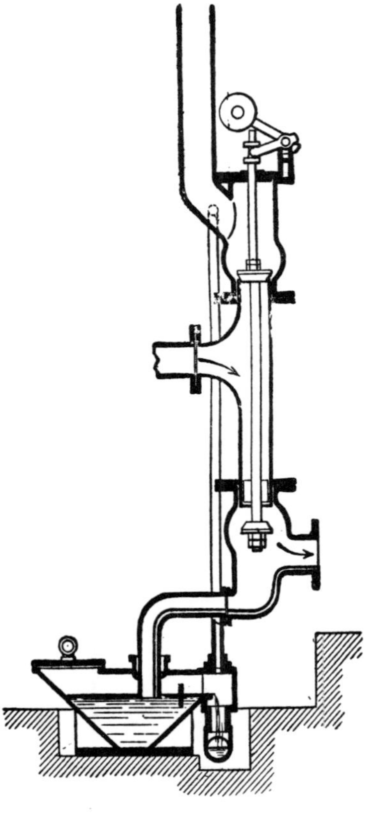

FIG. 113.—Otto Deutz flue.

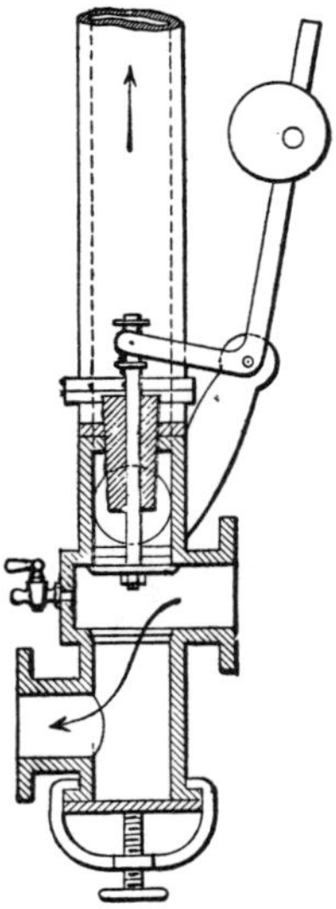

FIG. 114.—Benz flue.

cock (Bollinckx, Fig. 111), a three-way cock, a set of cocks, or, still better, a double valve, as in the Winterthur apparatus (Fig. 112) and the Deutz apparatus (Fig. 113). A double seated valve is also used, as is the case in the Benz generator (Fig. 114).

Vaporizer-Preheaters.—As has been stated before, there are vaporizers internal or external, relatively to the generator.

Internal Vaporizers.—The Deutz apparatus (Fig. 106), for example, consists of an annular cast-iron tank mounted above the retort of the generator.

The hot gases given off by the burning fuel travel around this tank and vaporize the water which it contains. The air drawn in by the suction of the engine enters through an opening located above the tank, travels over the surface of the water which is being vaporized, and thus laden with steam passes to the ash-pit.

The tank in question is supplied with water by means of a cock having a sight feed, located on the outside, and the level is kept constant by means of an overflow tube leading to the ash-pit. It is well to bend this tube and to place a funnel on its lower member. The amount of overflow may thus be regulated.

These vaporizers are simple and take up little room; but they are open to the apparently well-founded objection that they heat up slowly and require a considerable time to produce the steam necessary to enrich the gas, this being due to the relatively large mass of cast-iron and the amount of water contained therein.

The Pierson vaporizer (Fig. 101) and the Chavanon vaporizer (Fig. 115) both consist of an annular tank forming the base of the generator. Steam is formed near the outlet of the ashes, which, as has been described above, leads to the outer air. The development of steam is regulated by mechanical means controlled by the suction of the engine.

External Vaporizers.—External vaporizers are gen-

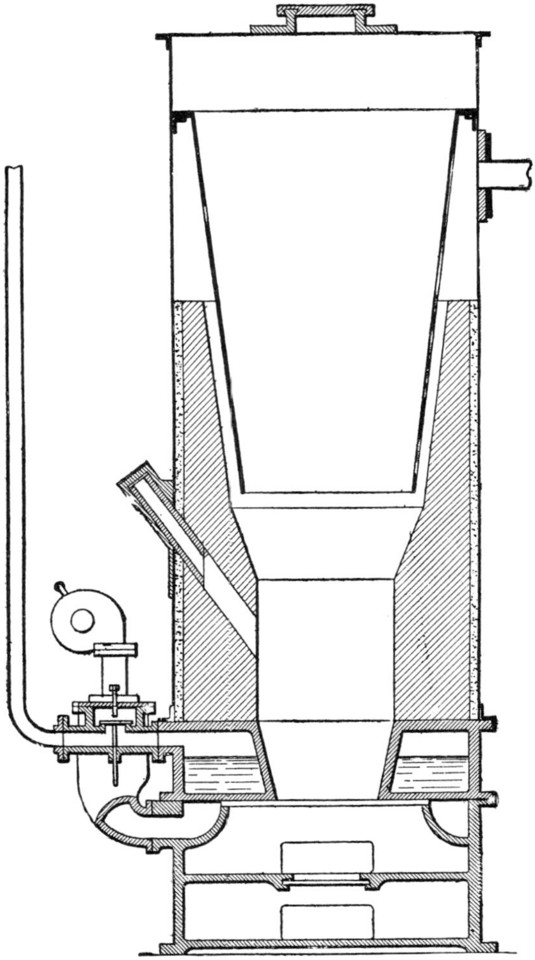

FIG. 115.—Chavanon producer.

erally formed by a cylinder with partitions constituting two series of chambers. In one of these the hot gases from the generator travel, and in the others the water to be vaporized is contained.

Tubular Vaporizers.—Different types of tubular vaporizers are manufactured. The vaporizer with a series of tubes, as in Taylor's apparatus (Fig. 116),

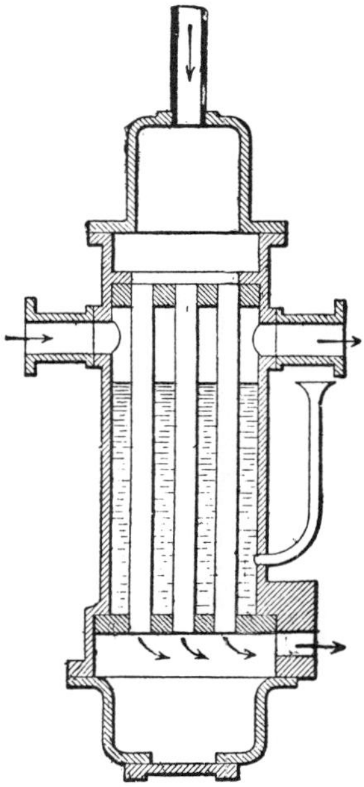

FIG. 116.—Taylor vaporizer.

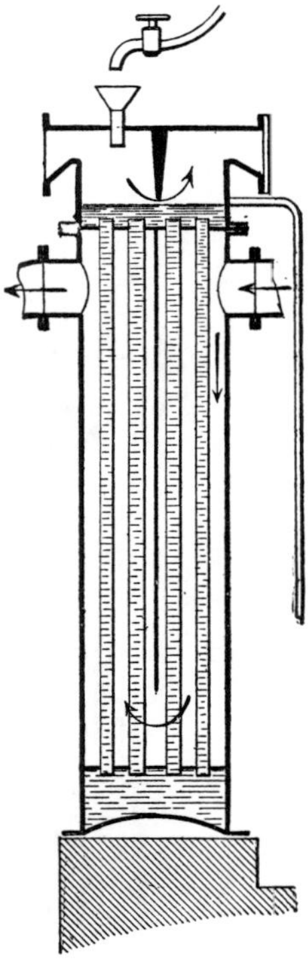

FIG. 117.—Deutz vaporizer.

Deutz's old model (Fig. 117), or with single tube like Pintsch's generator (Fig. 118), is formed by three compartments separated by two tube sheets or by plates which are connected by tubes.

In some cases the gases pass within the tubes, while the water to be vaporized surrounds them; as in the

Pintsch apparatus (Fig. 118), and Taylor apparatus (Fig. 116), Benz (Fig. 119), and Koerting generators (Fig. 120).

In other cases, the water lies inside and the gas out-

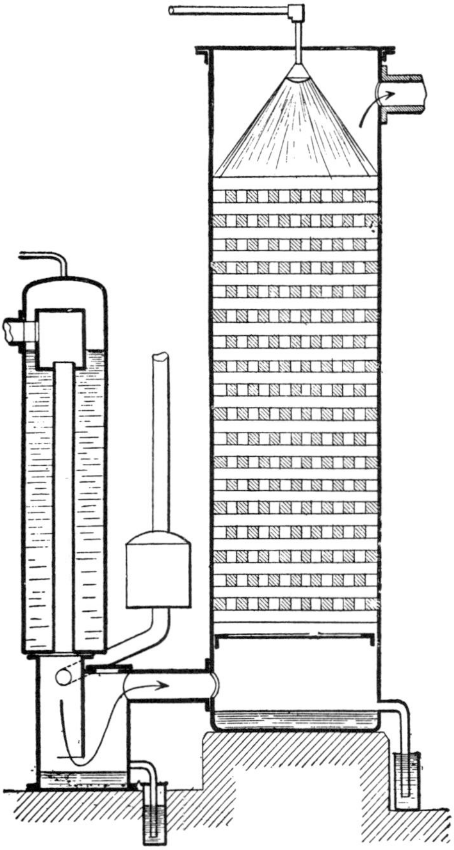

FIG. 118.—Pintsch vaporizer and scrubber.

side. In this latter case, a longitudinal baffle is employed to compel the gases to heat the tubes in their whole length, as in the Deutz producer (Fig. 117). In a general way it may be said that such a series of tubes presents the disadvantage of becoming clogged up rapidly by the deposit of lime salts contained in water.

If the set of tubes consists of fire-tubes, the deposit will form on the outer surface, that is, on a portion not accessible for cleaning. From this point of view,

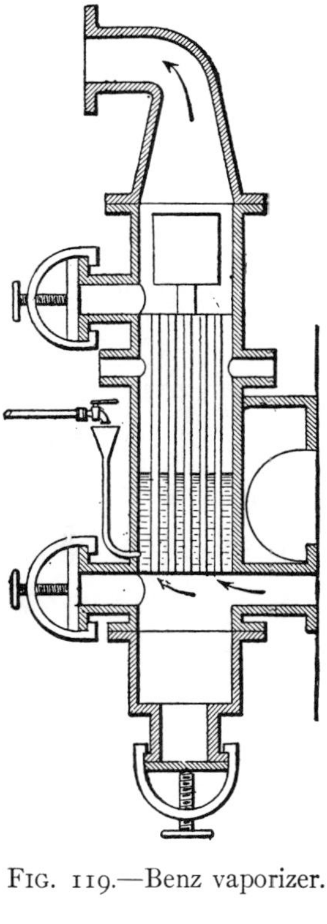

FIG. 119.—Benz vaporizer.

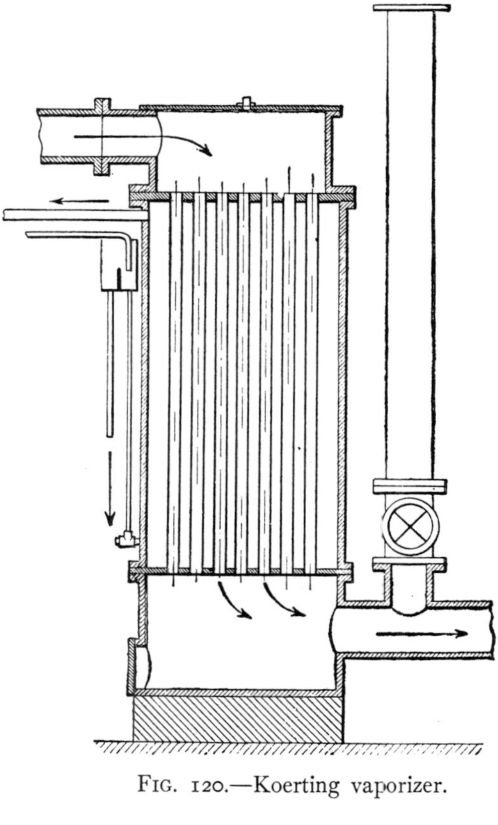

FIG. 120.—Koerting vaporizer.

water-tubes are preferable, as they allow the deposit or scale to be removed through the tubular heads or plates. On the other hand, such water-tubes have the drawback that their exterior surfaces are readily covered with pitch and soot. The tubular vaporizers of the

Field type (Bollinckx, Fig. 98) are composed of a single sheet-iron tube or shell, in which the tubes are arranged, dipping into a chamber through which the hot gases pass. This arrangement insures a rapid production of steam, but the Field tubes are even more liable than the others to become covered with deposits.

It will be seen that these types of vaporizers should all present the following features: easy access, small quantity of the body of water undergoing vaporization, and large heating surface with small volume.

The use of copper or brass tubes should be strictly avoided, as they would be quickly corroded by the action of the ammonia and hydrogen sulphide contained in the gas.

Partition Vaporizers.—Partition vaporizers comprise a cylindrical shell, generally made of cast-iron and having a double wall in which the water to be vaporized circulates. The gas coming from the generator passes into the central portion, where it comes in contact with a hollow baffle, also containing water (Wiedenfeld, Fig. 121). Vaporizers of this kind are strong, simple, and easily cleaned.

Operation of the Vaporizers.—The general purpose of vaporizers, whatever their construction may be, is to produce steam under atmospheric pressure, by utilizing the heat of the generator gases immediately after their production, or, as in the Chavanon system, by utilizing the heat radiated from the furnace.

The air drawn by the engine through the generator generally passes through the vaporizers and becomes laden with a certain amount of steam which it carries along. The amount thus taken up depends chiefly upon the temperature and the amount of gases coming from the generator, so that the greater the amount drawn into

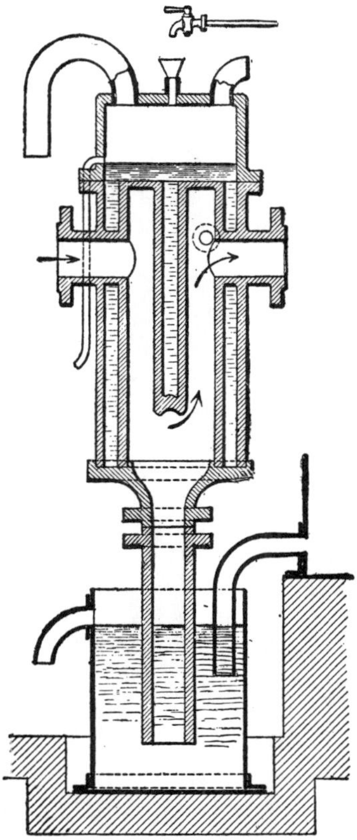

FIG. 121.—Wiedenfeld vaporizer.

the engine, the more energetic will the vaporization be, and the richer the gas will become. It will be understood that when a generator is working at its maximum production, the interior temperature is highest and most favorable to the decomposition of the largest amount of steam.

It follows that with the very simple vaporizers which have been reviewed, a practically automatic regulation is obtained. However, some manufacturers have deemed it advisable to regulate the amount of steam more accurately, and to make it exactly proportionate to the power developed by the motor. Thus in the Winterthur gas-producer (Figs. 92 and 112) the manufacturers have omitted the vaporizer proper, and use instead an air-heater and a super-heater for air and steam.

The heater is formed by a cast-iron box having two compartments, through one of which the hot gases from the generator pass, while in the other the air intended to support combustion travels. At the inlet of the super-heater a pipe terminates, which feeds, drop by drop, water supplied by a feed device to be described presently. This water is vaporized immediately upon contact with the wall of the super-heater and is carried along with the air contained in it.

The super-heater comprises a hollow ring-shaped cast-iron piece arranged in the chamber of the generator, in which the gases are developed, and is thus heated to a high temperature. The mixture of air and steam circulates in this super-heater before traveling to the ash-pit.

The feeder of the Winterthur gas-generator (Fig. 122) is composed of a receptacle having the shape of a tank or basin containing water and located below a closed cylindrical box. In this box a piston moves, which is provided at its lower end with a needle-valve.

The upper portion of the box communicates with the gas-suction pipe through a small tube. At each suction stroke of the engine, according to the force of the suction, the needle-valve piston rises more or less and thus allows a variable amount of water to pass.

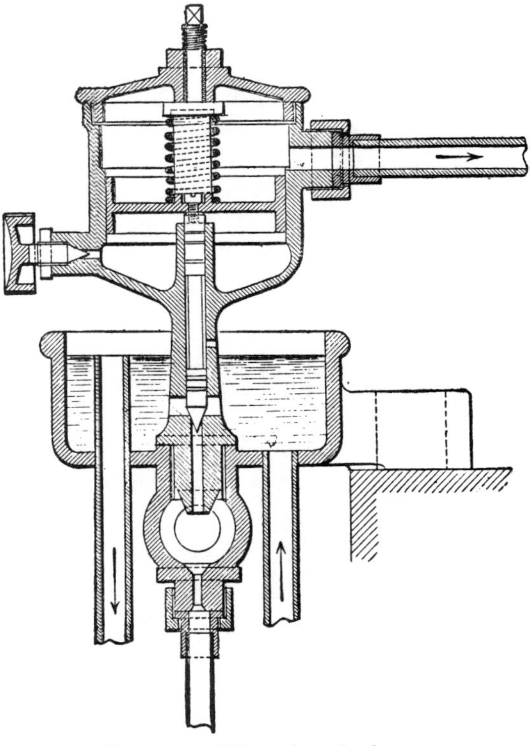

FIG. 122.—Winterthur feeders.

This apparatus—and all those based on the same principle—presents the advantage of proportioning the amount of water to the work of the engine; but in view of its rather sensitive operation it must be kept in perfect repair and carefully watched. Obviously, should the water contain impurities, the needle-valve will bind or the orifices will be obstructed, and thus the feeding of the water will be interrupted. This will not only result in the production of a poorer gas, but will lead to greater wear of the grates, which in this case are not sufficiently cooled by the introduction of steam.

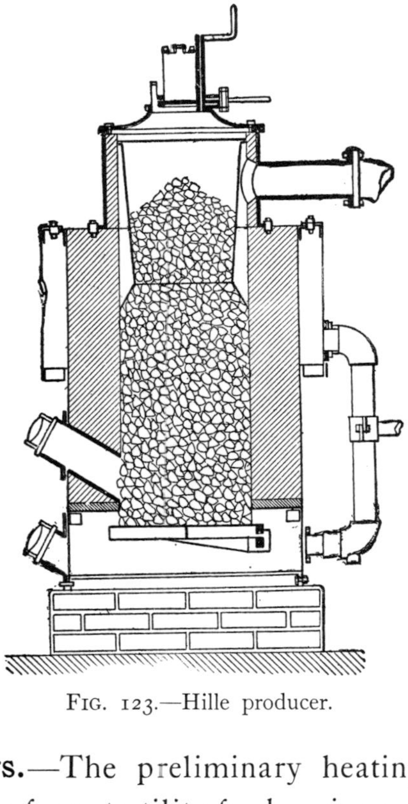

FIG. 123.—Hille producer.

Air-Heaters.—The preliminary heating of the air appears to be of great utility for keeping up a good fire. This heating is very easily accomplished, and is generally effected by utilizing a portion of the waste heat of the gases, a procedure which also has the advantage of cooling the gases before they pass through the washing apparatus.

The heating of the air for supporting combustion takes place either before the addition of steam (Hille's generator, Fig. 123), or after the mixture as in Wiedenfeld's apparatus (Fig. 95). In the first case, the air passes through a sheet-iron shell concentric with the basin of the generator, is there heated by the radiated heat, and is conveyed to the ash-pit by a tube into which leads the steam-supply pipe extended from the vaporizer. In the second type of heater, the mixture of air and steam is super-heated during its passage through an annular piece arranged in the ash-pit of the generator.

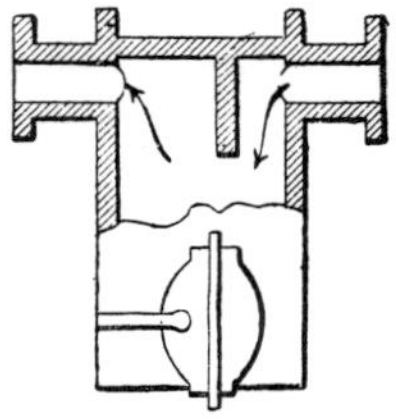

FIG. 124.—Benz dust-collector.

Dust-Collectors. — Dust-collectors are generally placed between the generator and the scrubber or washer. They may be formed of baffle-board arrangements against which the gases laden with dust impinge, causing the dust to be thrown down into a box provided with a cleaning opening (Benz, Fig. 124, and Pintsch, Fig. 118).

Some collectors are formed either by the vaporizer itself, terminating at its base in a tube which dips into water and forms a water-seal, as in the Wiedenfeld generator (Fig. 121), or by a water-chamber into which the gas-supply tube slightly dips (Bollinckx, Fig. 111).

With this arrangement, the gas will bubble through the water and will be partly freed of the dust suspended in it. These water-chambers are generally fed by the overflow from the spray of the scrubber. There is thus produced a continuous circulation by which the dust, in the form of slime, is carried toward the waste-pipe or sewer.

Cooler, Washer, Scrubber.—Some manufacturers cool the gas in a tower with water circulation. Most manufacturers, however, simply cool the gas in the washer or scrubber. This apparatus comprises a cylindrical body of sheet-iron or cast-iron formed of two compartments separated by a wooden or iron grate or perforated partition. The upper compartment up to a certain level contains either coke, glass balls, stones, pieces of wood, and the like. The top of the compartment is provided with a water supply in the nature of a sprinkler or spray nozzle. The lower compartment of the scrubber serves to collect the wash-water which has passed through the substance filling the tower. An overflow in the shape of a siphon, provided with a water seal, carries the water to the waste-pipe either directly or after it has first passed through the dust collector.

The gas drawn in enters the washer in the lower compartment either above the water level (Deutz, Fig. 125; Winterthur, Fig. 126), or through an elbow which dips slightly into the water (Benz, Fig. 127; Fichet and Heurtey producer, Fig. 128).

The gas passes through the grate or partition which supports the material filling the tower, and travels through the interstices in a direction opposite to that of the water falling from the top. Under these conditions, the gas is cooled, gives up the ammonia and the dust

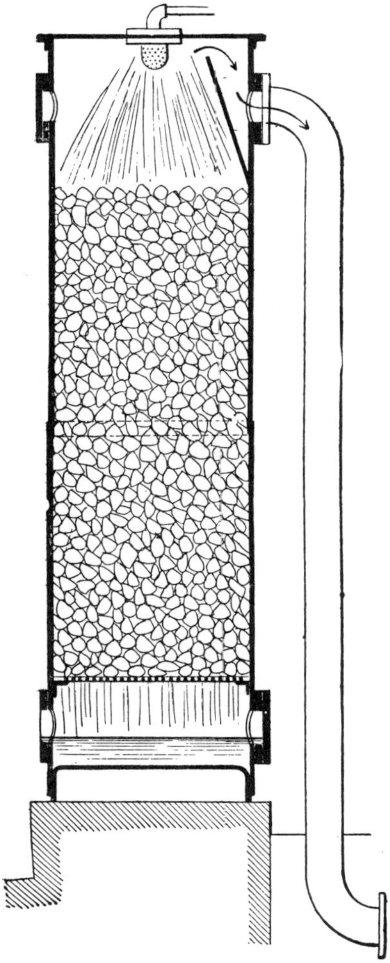

FIG. 125.—Otto Deutz scrubber.

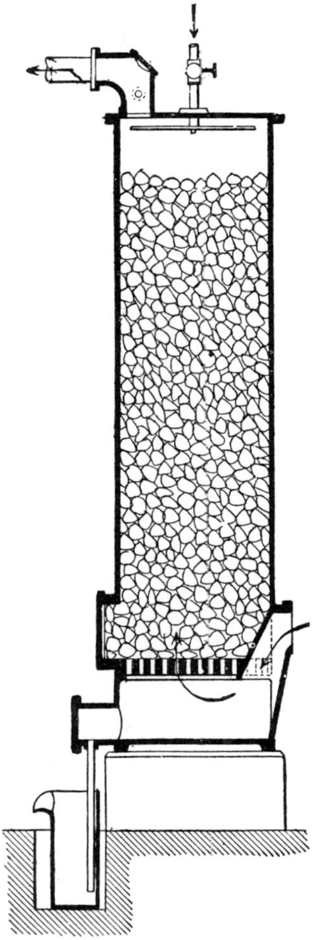

FIG. 126.—Winterthur scrubber.

which it may still contain in suspension, and is conveyed to the engine either directly or after passing through certain purifiers. Care should be taken to place the pieces of most regular shape along the walls, so that the unevenness of their surfaces may not form upward

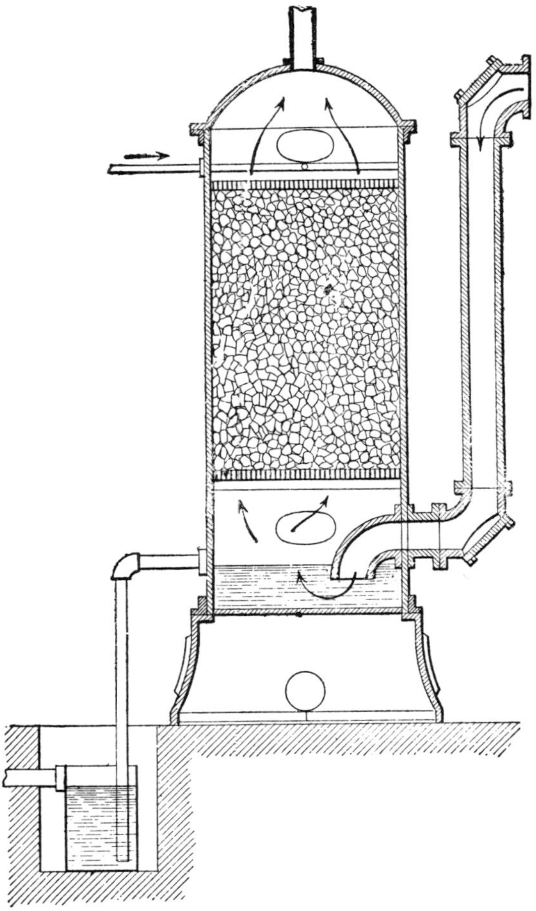

FIG. 127.—Benz scrubber.

channels along the shell, through which channels the gas could pass without meeting the wash-water.

The material most commonly employed in washers is coke in pieces of from 2½ to 3½ inches in size. This material is cheap and is very well suited for retaining the impurities of the gas. The largest pieces of coke should be placed at the bottom of the washer, and smaller pieces should form at the top a layer from 6 to

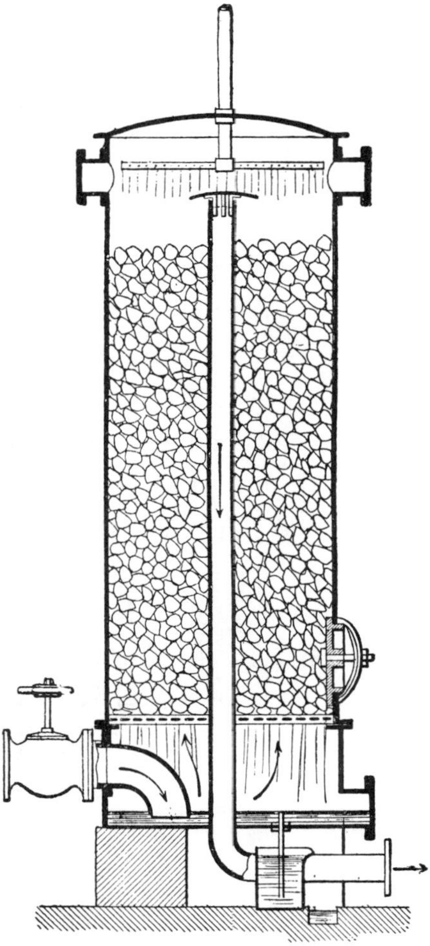

FIG. 128.—Fichet-Heurtey scrubber.

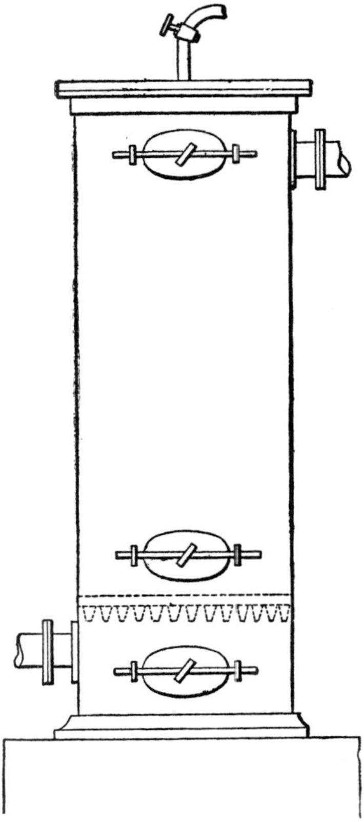

FIG. 129.—Scrubber-doors.

8 inches deep. In this manner the water is distributed more evenly and the gas is more thoroughly washed. Blast-furnace coke is best suited for this washing, as it is more porous and less brittle than gas-works coke. It

is advisable to put a baffle-board in front of the gas outlet to reduce the carrying along of water in the conduits.

The tower of the washer should be provided with three openings having air-tight closures, easily fastened by screws (Fig. 129). One of the openings is located in the lower compartment, slightly above the water level, to allow the deposits to be removed and to permit the cleaning of the orifice of the gas-supply tube, which is particularly liable to be obstructed. The second opening is placed above the grating which supports the filtering material. The third opening is provided on the top of the apparatus to permit the examination and cleaning of the water feed device and the gas outlet without the necessity of taking the lid of the washer apart, the joint of which is kept tight with difficulty. The two openings last mentioned also serve for introducing and removing the filtering material.

Purifying Apparatus.—In some cases, where it is necessary to have very clean gas or where coal is employed which is softer than anthracite coal, and which therefore produces an appreciable amount of tar, supplementary purifying means must be employed. The apparatus for this purpose may, like the washers, be based upon a physical action or upon a chemical action. The physical action has for its purpose chiefly to retain the pitch and the dust which may have passed through the washer.

This is accomplished by means of sawdust or wood shavings arranged in a thin layer and capable of filtering the gas without opposing too great a resistance to its passage. These materials are spread on one or more shelves superposed to form successive compartments in a box closed in an air-tight manner by an ordinary lid or a water seal cover (Pintsch, Fig. 130; Fichet and Heurtey, Fig. 131). It may be well to point out that the presence of the water carried along will, in the end, destroy the efficiency of the precipitated materials,

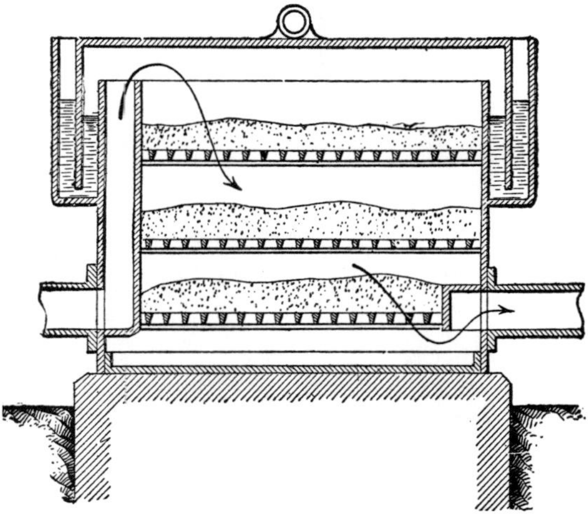

FIG. 130.—Pintsch purifier.

because they swell up and cease to be permeable to the gas. These materials must therefore be renewed rather frequently. To obviate this drawback, vegetable moss may be employed, which is much less affected by moisture than most filters and keeps its spongy condition for a long time.

The chemical action has for its chief object to rid the gas of the carbonic acid and the hydrogen sulphide which certain fuels give off in appreciable amounts.

The purifying material, in this case, is formed either by a mixture of hydrate of lime and natural iron oxide, or by the so-called Laming mass, which consists of iron sulphide, slaked lime, and sawdust, which last serves

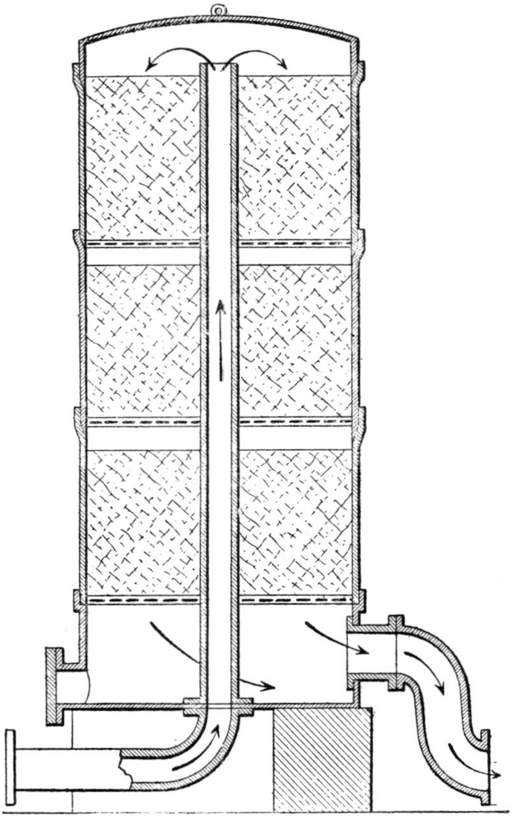

FIG. 131.—Fichet-Heurtey purifier.

the purpose of rendering the material looser and more permeable to the gas. The Laming mass as well as other purifying materials will become exhausted in the course of chemical reactions. It can be regenerated merely by exposure to the air.

Gas-Holders.—The purifiers by themselves constitute, to a certain extent, storage chambers for the gas before it is supplied to the engine; but in plants for the generation of gas without purifiers it is advisable to provide a gas-holder on the suction conduit near the engine.

In order to save floor space the gas-holder may be

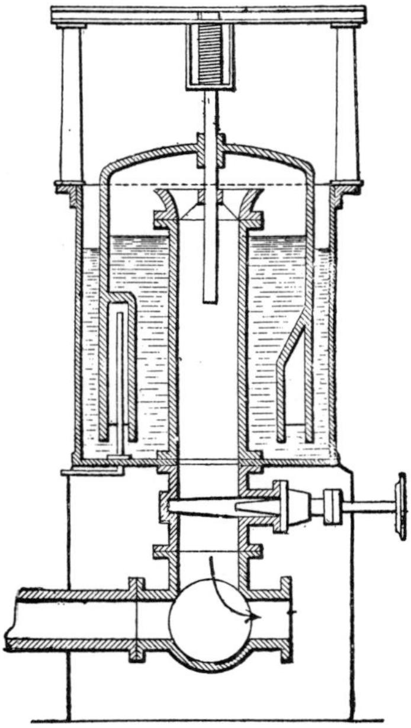

FIG. 132.—Pintsch regulating-bell.

placed in the basement. Preferably the capacity of the holder should be at least from 3 to 4 times the volume of the engine-cylinder. The holder should also be provided with a drain-cock and with a hand-hole located at some accessible point, so that the slimes and pitch which tend to accumulate in the holder can be removed. In some cases the gas-holder is formed by a small regulating bell, the function of which is to insure a uniform pressure. This bell is emptied during the suction period and is filled during the three suc-

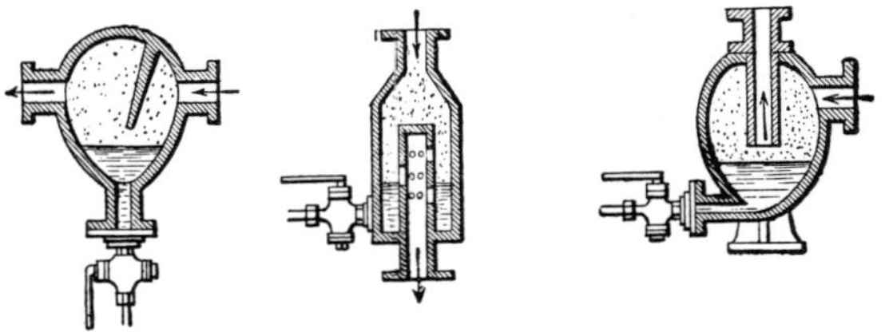

FIG. 133.—Types of gas-driers.

ceeding periods of compression, explosion, and exhaust (Pintsch, Fig. 132).

Drier.—Sometimes, toward the end of a producer-gas pipe, a drier is located for the purpose of keeping back the water carried along, the drier being similar to that employed in steam conduits. It will, of course, be

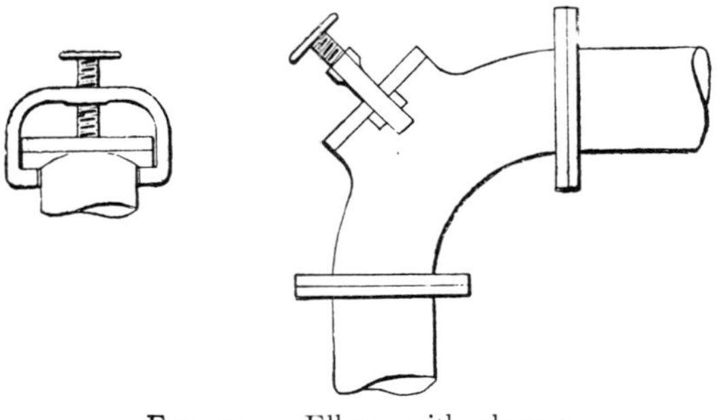

FIG. 134.—Elbow with closure.

understood that such driers are useful only in plants having no purifiers (Fig. 133). The employment of the drier is advisable to prevent the entrance of moist gas into the cylinder and the condensation of moisture on the electric igniter.

Pipes.—The pipes connecting the several parts of

a gas-producing plant should be disposed with particular care to insure tightness and cleanliness. It should be borne in mind that the gas is under a pressure below that of the atmosphere, and that the least leakage will cause the entrance of air, which will impair the quality of the gas. The greatest care should therefore be taken in fitting the joints. These joints are numerous, because there are joints wherever tubes are connected with each other and with the apparatus. Furthermore, all elbows should be provided with covers held in place by a yoke and compression screw, this being done for the purpose of providing for the introduction of a brush or other implement to remove the dust and pitch (Fig. 134).

For conduits of small diameter the elbows with covers may be replaced with **T** connections, or connections provided with plugs.

Gas piping in the immediate neighborhood of the cock for admitting gas to the motor should be provided with a conduit of proper diameter leading to the open air and serving to clean the apparatus and to fill them, during the operation of the fan, with gas suitable for combustion. This conduit should be provided with a stop-cock. Test-cocks for the gas should be placed on the piping immediately beyond the vaporizers, the scrubber, and near the engine.

It will also be well to provide water-pressure gages before and after the scrubber to enable the attendant to ascertain the vacuum in the conduits and to adjust the running of the apparatus.

Purifying-Brush.—As an additional precaution against the carrying of tar to the engine, metallic brushes are often employed, these brushes being spiral in form and enclosed in a cast-iron box interposed in the gas-supply pipe immediately after the engine. The

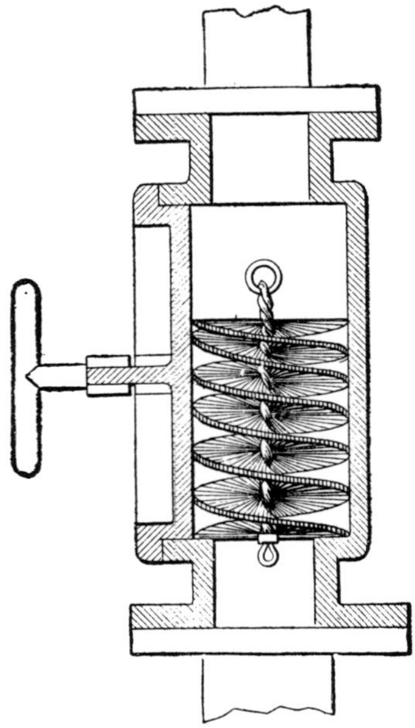

FIG. 135.—Metal purifying-brush.

gas will be broken up into streams by the obstacles formed by these brushes and will be freed of the suspended tar (Fig. 135). These brushes should be carefully cleaned at regular intervals. The best way of doing this is to drop them into kerosene or some other suitable solvent.

CONDITIONS OF PERFECT OPERATION OF GAS-PRODUCERS

These conditions depend upon the workmanship or upon the system of the plant, on the care with which it has been erected, on the nature of the fuel, on the condition of preservation of the apparatus, and upon the manner in which the producers have been working.

Workmanship and System.—The workmanship itself, which term is meant to include the choice of materials and the way they have been worked, presents no difficulty. The producers which we have discussed are very simple and offer absolutely no difficulties in their mechanical execution. As regards the system, however, especially with respect to the relative dimensions of the elements, it does not seem so far that it is possible to indicate any principle or rule capable of a rigid general application. It must be taken into account that the use of suction gas-generators has become general only in the last three or four years; the problem has therefore scarcely been adequately solved. However, some hints may be given on this subject.

Generator.—In regard to the generator, it is possible to deduce from the best existing plants the dimensions to be given to the generator relatively to those of the engine to be supplied, upon the assumption that the engine is single-acting and runs at a normal speed of from 160 to 230 revolutions per minute. The essential portion of the generator which contributes to the production of a proper gas is that which corresponds with the combustion zone. To this portion a cross-section is given varying in size between one-half and one-quarter of the surface of the engine-piston, sometimes between one-half and nine-tenths of this surface, according to the nature and the size of the fuel that is used. With small apparatus, however, ranging from 5 to 15 horse-power, the size of the base cannot be reduced below a certain limit, since otherwise the sinking of the fuel will be prevented. This danger always exists in small generators and renders their operation rather uncertain, such uncertainty being also due to the influence of the walls. It is to be noted that most modern generators are rather too large than otherwise.

Many manufacturers of no wide experience have been led to make their apparatus rather large so as to insure a more plentiful production of gas. As a matter of fact, the fire in such apparatus is liable to be extinguished when the combustion is not very active. If the principles of the formation of gas in suction-generators be kept in mind, it is evident that the gas developed is the richer the "hotter" the operation of the apparatus. Such operation also permits the decomposition of the hydrogen and carbon monoxide.

The "hot" operation of a generator is accomplished best with active combustion; and since this is a function of the rapidity with which the air is fed, it obviously is advantageous to reduce the area of the air-passage to a minimum as far as allowed by the amount of fuel to be treated. As to the height of the fuel in use in the apparatus, this varies as a rule between 4 and 5 times the diameter at the base.

Vaporizer.—The size of the vaporizer varies materially according to its type. No hard-and-fast rule can therefore be adopted for determining its heating surface; but this surface should in all cases be sufficient to vaporize under atmospheric pressure from .66 to .83 pounds of water per pound of anthracite coal consumed in the generator.

Scrubber.—For the scrubbers, the following dimensions may be deduced from constructions now used by standard manufacturers.

The volume of a scrubber is generally from six to eight times the anthracite capacity of the generator. A height of from three to four times the diameter is considered sufficient in most cases. It should be understood that in this height is included the water-pan chamber located below the partition or grate, and the upper chamber through which the gas escapes. The height of these two chambers depends necessarily upon the arrangement used for leading the gas to the lower portion of the washer and for the distribution of wash-water at the top.

Assembling the Plant.—The author has insisted strongly on the necessity of having all the apparatus and pipe connections perfectly tight. In order to ascertain if there is any leakage, the following procedure may be adopted:

When starting the fire by means of wood, straw, or other fuel producing smoke, instead of allowing this smoke to escape through the flue during the operation of the fan, it may be caused to escape through the cock which generally admits the gas to the motor, the cock being opened for this purpose. The damper in the outlet flue is closed. In this manner the smoke will fill all the apparatus and connecting pipes under a certain pressure and will escape through any cracks, the presence of which will thus be revealed.

Another test, which is made during the ordinary operation of the generator, consists in passing a lighted candle along the joints; if there is any leakage, this will be shown by a deviation of the flame from a vertical position.

Fuel.—We have discussed the subject of fuel in a preceding chapter (Chapter XIII) and have indicated the conditions to be fulfilled by low grade or anthracite coal best adapted for use in suction gas-generators. It may here be added that the coal used in the generator should be as dry as possible and in pieces of from ½ inch to 1 inch. Very small pieces, and particularly coal dust, are injurious and should be removed by preliminary screening as far as possible. Screened coal is thrown in with an ordinary grate shovel.

How to Keep the Plant in Good Condition.—In regard to the generator, apart from the cleaning of the grate and of the ash-pit, which may be done during operation, it is necessary to empty the apparatus entirely once a week, if possible, in order to break off the clinkers adhering to the retort. These clinkers destroy the refractory lining, form rough projections interfering with the downward movement of the fuel, bring about the formation of arches, and reduce the effective area of the retort. At the time of this cleaning, tests are also made as to the tightness of the doors of the combustion-chamber, of the charging-boxes, etc.

The vaporizer should be cleaned every week or every other week, according to the more or less bituminous character of the fuel and the greater or smaller content of lime in the water used. Lime deposits may be eliminated, or the salts may be precipitated in the form of non-adhering slimes, by introducing regularly a small amount of caustic postash or soda into the feed-water. If the deposits or incrustations are very tenacious, the use of a dilute solution of hydrochloric acid may be resorted to. Tar which may adhere to the conduits, pipes or gas passages, is best removed while the apparatus is still hot, or a solvent may be employed, such as kerosene, turpentine, etc. The connections between the vaporizer and the scrubber are particularly liable to become obstructed by the accumulation of tar or dust carried along by the gas.

It is advisable to examine the several parts of the plant once or twice a week by opening the covers or the cleaning-plugs.

The lower compartment of the washer keeps back the greater part of the dust which has not been retained in collectors or boxes provided especially for this purpose. The dust takes the form of slime, and, in some arrangements of apparatus, tends to clog up the overflow pipe, thus arresting the passage of gas and causing the engine to stop. This portion of the washer should be thoroughly cleaned once or twice a month.

If very hard blast-furnace coke is used in the washer, it may be kept in use for over a year without requiring removal. In order to free the purifying materials from dust and lime sediments carried along by the wash-water, it is well to let the wash-water flow as abundantly as possible for about a half-hour at least once a month. At the time of renewing the purifying material the precautions indicated in the section dealing with these matters should be observed, and care should be taken to have shelves or gratings on which the material is supported in layers not too thick, so as to avoid any resistance to the passage of the gas.

In a general way it is advisable to test the drain-cocks on the several apparatus daily, and to keep them in perfect condition. If, when open, one of these cocks does not discharge any gas, water, or steam, a wire should be introduced into the bore to make sure it is not clogged up.

Care of the Apparatus.—Each producer-gas plant will require special instructions for running it, according to the system, the construction, and the size of the plant. Such instructions are generally furnished by the manufacturer. However, there are some general rules which are common to the majority of suction gas-producers, and these will here be enumerated.

Starting the Fire for the Gas Generator.—This

operation calls for the presence of the engineer of the plant and an assistant. The proper procedure is as follows:

First: Open the doors of the furnace and of the ash-pit. Then open the outlet flue and make sure that the grate of the generator is clear of ashes and clinkers. It should also be seen to that the parts of the charging-box work well and that the joints are tight.

Second: Ascertain whether there is the proper amount of water in the vaporizer, in the scrubber, etc., and that the feed works properly.

Third: Through the door of the combustion-chamber introduce straw, wood shavings, cotton waste, etc.; light them and fill the generator with dry wood up to one-quarter or one-half of its height; then add a few pailfuls of coal.

Fourth: Close the doors of the ash-pit and of the combustion-chamber and start the draft by means of the fan. As soon as the draft is started, it must be kept up without interruption until the engine begins to run, which may be ten or twenty minutes after lighting the fire.

Fifth: After the draft has been continued for a few minutes, the coal becomes sufficiently incandescent to start the production of gas, which may be ascertained by trying to light the gas at the test-cock near the generator. Then the opening in the outlet flue is half closed for the purpose of producing pressure in the apparatus.

Sixth: Open the outlet flue adjacent to the engine for

the purpose of purging the apparatus and the conduits of the air which they contain until the gas may be lighted at the test-cock placed near the motor.

Seventh: Adjust the normal outflow of wash-water for the scrubber.

Eighth: As soon as the gas burns continuously at the test-cock with an orange-colored flame the engine may be started.

The gas at first burns with a blue flame; this color indicates that it contains a certain amount of air. The opening of the test-cock should be so regulated as to reduce the outlet pressure of the gas sufficiently to prevent the flame from going out. During the production of the draft, as well as during the ordinary running of the plant, the filling of the apparatus with fuel should be done with care to prevent explosions of gas due to the entrance of air. Particular care should be taken never to open at the same time the lid of the charging-box and the device, be it a cock, valve, or damper, which controls the connection of the charging-box with the generator. All the operations which have been mentioned above should be carried out as quickly as possible.

STARTING THE ENGINE

The manner of starting the engine depends on the type of the engine and on the starting device with which it is provided, as we have already explained in connection with engines working with gas from city mains.

It is, however, important for the production of a good explosive mixture to regulate the amount of air supplied to the engine according to the quality of the gas employed. It is advisable to continue the operation of the fan until several explosions have taken place in the cylinder and the engine has acquired a certain speed so as to be able to draw in the normal amount of gas.

Naturally the gas-outlet tube near the admission-cock should be closed after starting the engine, as well as the opening in the outlet flue of the generator. When the motor is running properly, the amount of water fed to the vaporizer and overflowing to the ash-pit is properly adjusted. The generator is then filled up to the level indicated by the manufacturer.

Care of the Generator during Operation.—As soon as the apparatus is running under normal conditions, it presents the advantage of requiring only very slight supervision and very little manual tending. The supervision consists:

First: In regulating and keeping up a proper feed of water to the vaporizer.

Second: In seeing to it that in apparatus provided with an overflow leading to the ash-pit, the water should flow constantly but without exceeding the proper amount.

Third: In keeping down temperature in the scrubber by properly regulating the feed of the wash-water. This apparatus may be slightly warm at its lower part, but must be quite cold at the top.

The manual tending to be done is limited to the regular filling up of the generator with fuel and to the removal of ashes and clinkers. The charging is effected at regular intervals, which, according to the various types of anthracite-generators, vary from one to six hours. Charging the apparatus at short intervals entails unnecessary labor, while charging at too long intervals will often interfere with the uniform production of the gas.

It will be obvious that the amount of fuel introduced will be the larger, the greater the intervals between two fillings. This fuel is cold and contains between its particles a certain amount of air; furthermore, the layer of coal which covers the incandescent zone has become relatively thin. The excess of air impoverishes the gas, and the fresh fuel lowers the temperature of the mass undergoing combustion, so that again the gas in process of formation is weakened. Experience seems to show that as a rule it is best to fill up the generator at intervals of from two to three hours, according to the work done by the engine. It should be noted that the level of the fuel in the generator should not sink below the bottom of the feed-hopper.

The author wishes again to emphasize that in order to prevent the harmful entrance of air, the charging operations should be carried out as quickly as possible; and for this reason the fuel should be introduced not by means of the shovel, but by means of a pail, scuttle, or other appropriate receptacle.

Care should be taken to fill the charging box to its

upper edge and to adjust its cover accurately before operating the device which closes the feed-hopper (valve, cock).

The removal of the ashes and clinkers should be accomplished as infrequently as possible, since opening the doors of the ash-pit and of the combustion-chamber necessarily causes an inward suction of cold air which is harmful.

As a rule with generators employing anthracite coal, it is sufficient to empty the ash-pit twice daily; this should be preferably done during stoppages. However, the cleaning of the grate by means of a poker passed between the grate-bars or over them in order to bring about the falling of the ashes, should be attended to every two to four hours, according to the type of the generator and the nature of the fuel. In order that this cleaning may be done without opening the doors, the latter should be provided with apertures having closing devices.

This cleaning has for its chief object to allow the free passage of the air for supporting combustion and to keep the incandescent zone in the apparatus at the proper height. The accumulation of ashes and clinkers at the bottom of the retort will shift this zone upward and impair the quality of gas.

Stoppages and Cleaning.—After closing the gas-inlet to the engine, the damper in the gas-outlet flue of the generator should be opened and the cocks controlling the feed of water to the scrubber and to the vaporizer should be closed.

If it is desired to keep up the fire of the generator during the stoppage so as to be able to start again quickly, the ash-pit door should be opened so as to produce a natural draft which will maintain combustion. While the door is open, the clinkers which have accumulated above the grate may be removed, as they are much more easily taken off the grate when they are hot.

At least once a week the fire in the generator should be put out and the generator completely cleaned—that is, when ordinary fuel is employed. For this purpose, as soon as the apparatus is stopped, a portion of the incandescent fuel is withdrawn through the doors of the combustion-chamber, and the retort is allowed to cool before it is emptied entirely. Too sudden a cooling of the retort may injure its refractory lining. In order to prevent explosions caused by the entrance of air, the feed-hopper should remain hermetically closed during the removal of the incandescent fuel through the doors of the combustion-chamber.

If the apparatus is placed in a room poorly ventilated, the cleaning should be attended to by two men, so that one may assist the other in case he is overcome by the gas. In all cases there should be a strict prohibition against the use of any light having an exposed flame liable to set on fire the explosive mixtures which may be formed.

When the generator, after cooling, is completely open, the charging-box is taken apart, and, if necessary, the feed-hopper also; the grates are taken out, if necessary; and, by means of a poker inserted from above, the clinkers and slag adhering to the retort are broken off.

In the foregoing paragraphs the author has indicated how the several apparatus, such as the vaporizer, the washer, the conduits, etc., should be attended to and maintained in good working order.

CHAPTER XIV

OIL AND VOLATILE HYDROCARBON ENGINES

ALTHOUGH this book is devoted primarily to a discussion of street-gas and producer-gas engines employed in various industries, a few words on oil and volatile hydrocarbon engines may not be out of place.

Oil-engines are those which use ordinary petroleum as a fuel or illuminating oil of yellowish color, having a specific gravity varying from 0.800 to 0.820 at a temperature of 15 degrees C. (490 degrees F.), and boiling between 140 and 145 degrees C. (284 to 297 degrees F.). Volatile hydrocarbon engines are those which employ light oils obtained by distilling petroleum. These oils are colorless, have a specific gravity that varies from 0.680 to 0.720, and boil between 80 degrees and 115 degrees C. (176 to 257 degrees F.). Among these "essences," as they are called in Europe, may be mentioned benzine and alcohol.

In general appearance, and the way in which they are controlled, oil-engines differ but little from gas-engines. Their usual speed, however, is 20 to 30 per cent. greater than that of gas-engines. Except in some engines of the Diesel and Banki types, the compression does not exceed 43 to 71 pounds per square inch. In volatile hydrocarbon engines, on the other hand, the speed is very high, often running from 500 to 2,000 revolutions per minute, while the speed of gas or oil engines rarely exceeds 250 or 300 revolutions per minute.

Oil-Engines.—Oil-engines are employed chiefly in Russia and in America. Because of the high price of oil in other countries they are to be found only in small installations in country regions and are used mainly for driving locomobiles and launches. The improvements which have been made of late years in the construction of gas-engines supplied by suction gas-producers for small as well as for large powers, have hindered the general introduction of oil-engines.

The characteristic feature in the design of many of the oil-engines of the four-cycle type now in use (to which type we shall confine this discussion) is to be found in the controlling mechanism employed. The underlying principle of this mechanism lies not in acting upon the admission-valve, but in causing the governor to operate the exhaust-valve in such a manner that it is held open whenever the engine tends to exceed its normal speed. Some engines, however, are built on the principle of the gas-engine, with an admission-valve so controlled by the governor that it is open during normal operation and closed whenever the speed becomes excessive.

The necessity of producing a mixture of air and oil capable of being ignited in the engine-cylinder has led to the invention of various contrivances, which cannot be used if illuminating-gas or producer-gas be employed. These contrivances are the atomizer, the carbureter, the oil-pump, the air-pump, the oil-tank, and the oil-lamp. In some oil-engines all of the elements may be found, but for the purpose of simplifying the construction and of avoiding unnecessary complications, manufacturers devised arrangements which rendered it possible to discard some of them, particularly those of delicate construction and operation. It is not the intention of the author to enter into a detailed description of these various devices, since the limitations of this book would be considerably surpassed. The reader is referred to books on the oil-engine, published in the United States, England, and France.*

Most of the observations which have been made on the construction and installation of gas-engines, as well as the precautions which have been advised in the conduct of an engine, apply with equal force to oil-engines. It will therefore be unnecessary to recur to this phase of the subject so far as oil-engines are concerned. One point only should be insisted upon—the necessity of very frequently cleaning the valves and moving parts of the engine.

Illuminating-oil when burnt produces sooty deposits, particularly if combustion be incomplete, which deposits foul the various parts and cause premature ignitions and faulty operation.

* Hiscox, Gas and Oil Engines, Norman W. Henley Pub. Co., New York. Parsell and Weed, Gas and Oil Engines, 1900, Norman W. Henley Pub. Co., New York. Goldingham, 1900, Spon & Chamberlain, London. Dugald Clerk, 1897, Longmans, London. Grover, 1902, Heywood, Manchester. Aimé Witz, 1904, Barnard, Paris. H. Güldner, 1903, Springer, Berlin.

The use of oil in atomizers, carbureters, and lamps is accompanied with the employment of pipes and openings so small in cross-section that the slightest negligence is attended with the formation of partial obstructions that inevitably affect the operation of the engine.

Volatile Hydrocarbon Engines.—Only those engines will here be treated which have become of importance in the development of the automobile.

Some designers have attempted to employ the volatile hydrocarbon engine for industrial and agricultural purposes, and have devised electro-generator groups, hydraulic groups, and so-called " industrial combinations " in which belt and pulley transmission is employed. These applications in particular will here be rapidly reviewed.

The high speed at which engines of this class are driven renders it possible to operate a centrifugal pump directly and to mount both the engine and machine which it actuates on the same base. The hydrocarbon engine has the merit of being very light and of taking up but little room. Its cost is considerably less than that of an oil or producer-gas engine of corresponding power. On the other hand, its maintenance is much more expensive, and the hydrocarbons upon which it depends for fuel anything but cheap. Furthermore, the engines wear away rapidly, on account of their high speed. For this reason it is advisable to base calculations on a life of three to four years, while oil and gas engines may generally be considered to be still of service at the end of thirteen years. On the following page a comparison of costs for installation and maintenance is drawn between the oil and hydrocarbon engine on the basis of ten horse-power.

Comparative Costs.—A 10 horse-power oil-engine, in the matter of first cost of installation, is about 35 per cent. more expensive than a volatile hydrocarbon engine of equal power. On the other hand, the operating expenses of the oil-engine are less by 25 per cent. than they are for the volatile hydrocarbon engine.

The engines which are here discussed usually have their cylinders vertically arranged, as in steam-engines of the overhead cylinder type. The crank-shaft and the connecting-rods are enclosed in a hermetically sealed box filled with oil, so that the movement of the parts themselves ensures the liberal lubrication of the piston. The suction-valve is generally free, although latterly designers have shown a tendency to connect it with the cam-shaft, with the result that it has become possible to reduce the speed appreciably without stopping the engine. The carbureter is operated by the suction of the engine. If the fuel employed is alcohol, it must be heated.

Tests of High-speed Engines.—High-speed engines present various difficulties which must be contended with in controlling their operation. Their high speed renders it impossible to take indicator records as in the case of most industrial engines. Indicator cards, moreover, at best give but very crude data, which relate to each explosion cycle only, and which are therefore inadequate in determining the exact conditions of an engine's operation. Oil, benzine, and other so-called carbureted-air engines are particularly difficult to control because of many phenomena which cannot be recorded. In order to test the operation of high-speed engines, two different types of instruments are at present employed: the manograph and the continuous explosion recorder.

The Manograph.—The manograph, which is the invention of Hospitalier, is an optical instrument in which a series of closed diagrams are superimposed upon a polished mirror similar in form to Watt diagrams. Because the images persist in affecting the retina of the eye an absolutely continuous, but temporary, gleam is seen. Still, it is possible to obtain a photograph or a tracing of these diagrams.

The Continuous Explosion Recorder for High-speed Engines.—The author has devised an explosion and pressure recorder, which is mounted upon the explosion chamber to be tested and which communicates with the chamber through the medium of a cock *r* (Fig. 136). The instrument is somewhat similar in form to the ordinary indicator. Its record, however, is made on a paper tape which is continuously unwound. The cylinder *c* is provided with a piston *p*, about the stem of which a spring *s* is coiled. A clock train contained in the chamber *b* unwinds the strip of paper from the roll p' and draws it over the drum p', where the pencil *t* leaves its mark. The tape is then rewound on the spindle p'''. A small stylus or pencil *f* traces " the atmospheric line " on the paper as it passes over the drum p''. In order to obviate the binding of the piston p when subjected to the high temperature of

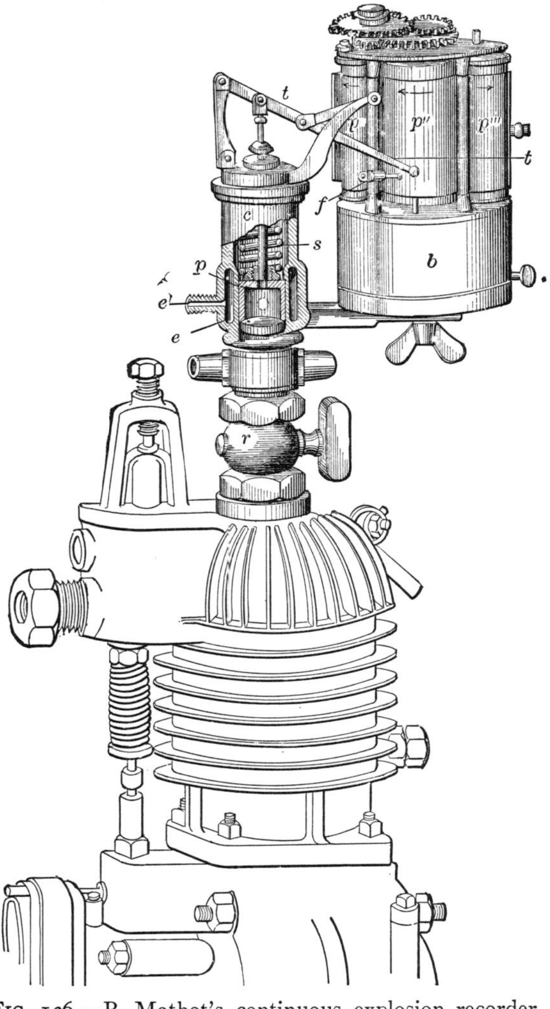

FIG. 136.—R. Mathot's continuous explosion recorder.

the explosions, the cylinder c is provided with a casing e in which water is circulated by means of a small rubber tube which fits over the nipple e'. This recorder analyzes with absolute precision the work of all engines, whatever may be their speed. It gives a continuous graphic record from which the number of explosions, together with the initial pressure of each, can be determined, and the order of their succession. Consequently the regularity or irregularity of the variations can be observed and traced to the secondary influences producing them, such as the section of the inlet and outlet valves and the sensitiveness of the governor. It renders it possible to estimate the resistance to suction and the back pressure due to expelling the burnt gases, the chief causes of loss in efficiency in high-speed engines. Furthermore, the influence of compression is markedly shown from the diagram obtained.

The recorder is mounted on the engine; its piston is driven back by each of the explosions to a height corresponding with their force; and the stylus or pencil controlled by the lever *t* records them side by side on the moving strip of paper. The speed with which this strip is unwound conforms with the number of revolutions of the engine to be tested, so that the records of the explosions are placed side by side clearly and legibly. Their succession indicates not only the number of explosions and of revolutions which occur in a given time, but also their regularity, the number of misfires. The atmospheric pressure of the explosions is measured by a scale connected with the recorder-spring. By employing a very weak spring which flexes at the bottom simply by the effect of the compression in the engine-cylinder, it is possible

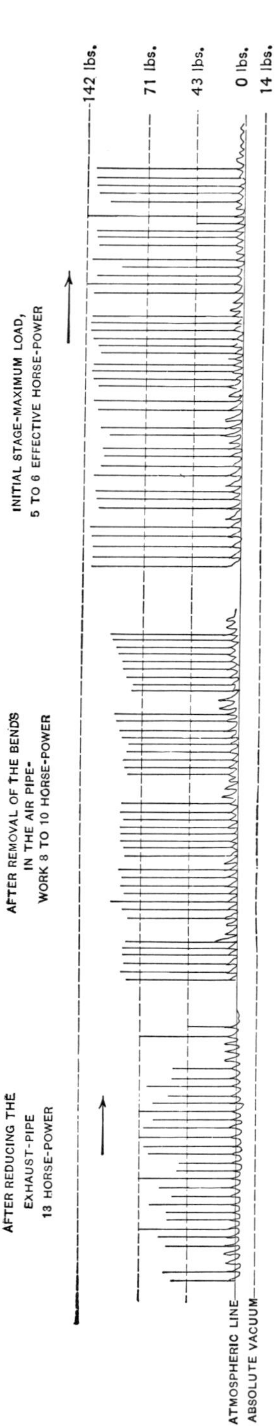

FIG. 137.—12 H. P. Oil-engine.

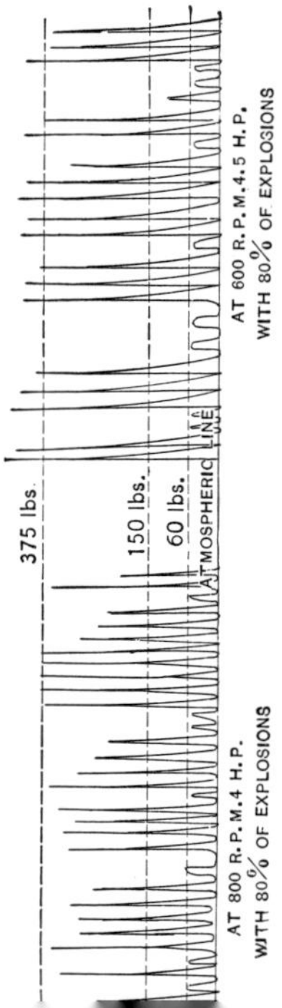

FIG. 138.—6 H. P. Volatile Hydrocarbon Engine.

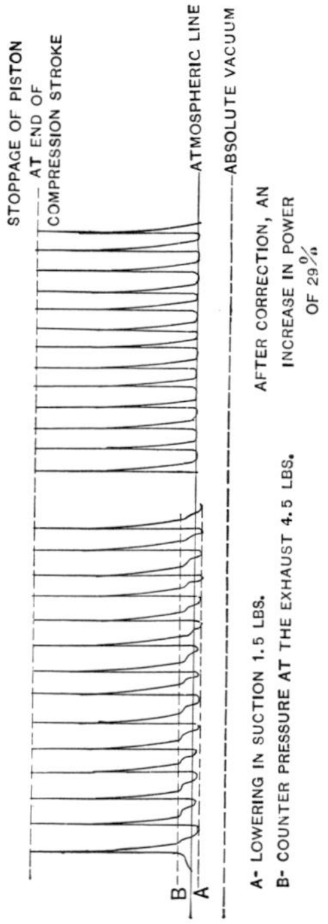

FIG. 139.—Effect of size of section and exhaust ports.

to ascertain the amount of the resistance to suction and to the exhaust. It is simply sufficient to compare the explosion record with the atmospheric line, traced by the stylus *f*. By means of this apparatus, and of the records which it furnishes, it is possible analytically to regulate the work of an engine, to ascertain the proportion of air, gas, or hydrocarbon, which produces the most powerful explosion, to regulate the compression, the speed, the time of ignition, the temperature, and the like (Figs. 137, 138 and 139).

In order to explain the manner of using this recorder several specimen diagrams are here given.

I. *Determination of the Amount of Compression.*—A spring of average power is employed, the total flexion of which corresponds almost with the maximum compression so as to obtain a curve of considerable amplitude. The engine is first revolved without producing explosions, driving it from the dynamo usually employed in shops, at the different speeds to be studied. The compression of the mixture varies in inverse ratio to the number of revolutions of the shaft, owing to the resistances which are set up in the pipes and the valves and which increase with the speed. The accompanying cut (Fig. 140) shows two distinct records taken in two different cases, namely:

A.—Speed of engine, 950 revolutions per minute; amount of compression, 68.9 pounds per square inch.

B.—Speed of engine, 1,500 revolutions per minute; amount of compression, 61 pounds per square inch, or 11.5 per cent. less.

II. *Determination of the Resistance to Suction and Exhaust.*—Influence of the tension of the spring of the suction valve and of the section of the pipe. Effect of

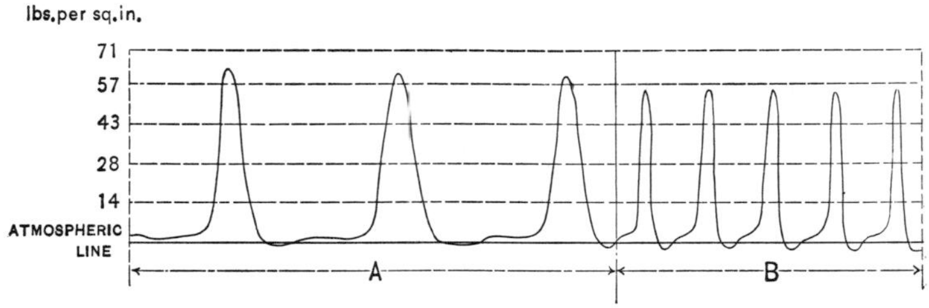

FIG. 140.

the section of the exhaust-valve and of the length and shape of the exhaust-pipe:

A very light spring is utilized, the travel of which is limited by a stop so as to obtain on a comparatively large scale the depressions and resistance respectively represented by the position of the corresponding curve, above or below the atmospheric line (Fig. 141).

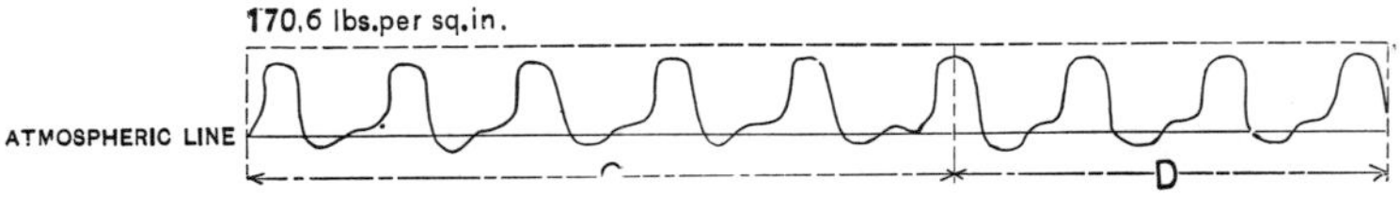

FIG. 141.

C.—Tension of the suction-valve: 2.9 pounds. Resistance to suction: $\frac{1}{7}$ of an atmosphere (2.7 pounds).

D.—Tension of the suction-valve: 2.17 pounds. Resistance to suction: $\frac{2}{7}$ of an atmosphere (5.4 pounds).

E.—A chest is used for the exhaust. Resistance to exhaust: $\frac{2}{7}$ of an atmosphere (5.4 pounds).

F.—The exhausted gases are discharged into the air,

the pipe and the chest being discarded. Resistance to the exhaust is zero (Fig. 142).

The depression graphically recorded is partly due to the inertia of the spring of the explosion-recorder,

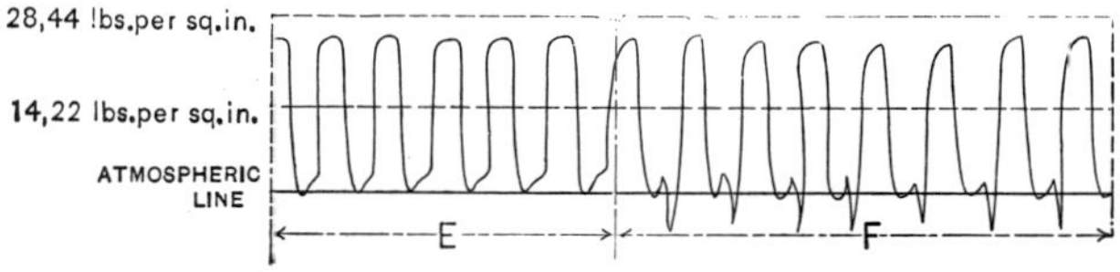

FIG. 142.

which spring expands suddenly when the exhaust is opened.

III. *Comparison of the Average Force of the Explosions by Means of Ordinates.*—A powerful spring is employed. The paper band or tape of the recorder is moved with a small velocity of translation so as to approximate as closely as possible the corresponding ordinates representing the explosions (Fig. 143).

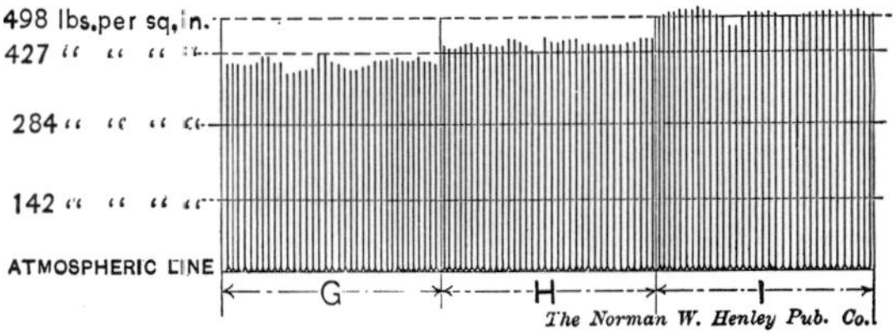

FIG. 143.

G.—Pure alcohol. Explosive force, 369.72 to 426.6 pounds per square inch.

H.—Carbureted alcohol. Explosive force, 397.6 to 510.8 pounds per square inch.

I.—Volatile hydrocarbon. Explosive force, 483.48 to 531.92 pounds per square inch.

IV. *Analysis of a Cycle by Means of Open Diagrams Representing the Four Periods.*—A powerful spring is employed, and the paper is moved with its maximum speed of translation. The four phases of the cycle are easily distinguished as they succeed one another graphically from right to left—in other words, in a direction opposite to that in which the paper is unwound. A diagram is made which reproduces exactly the values of the corresponding pressures at different points in the travel of the piston (Fig. 144). The periods of the

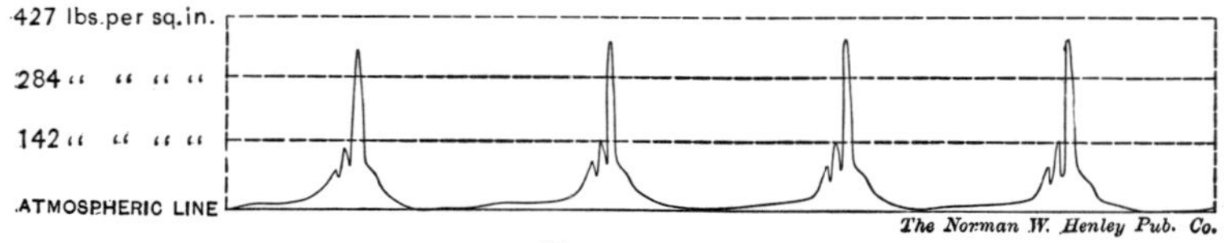

FIG. 144.

cycle are reproduced as faithfully as if the ordinary indicator which gives a closed curved diagram had been employed. There is no difficulty in reading the record, since the paper is not in any way connected with the engine-piston. Some attempts have been made to secure open diagrams in which the motion of translation given to the paper is controlled by the engine itself; but these apparatus as well as the ordinary indicators cannot be used when the speed of the engine exceeds 400 to 500 revolutions per minute.

J.—Speed, 1,200 revolutions; carbureted alcohol; average force of the explosions, 426.6 pounds per square inch. Average compression, 92.43 pounds per square inch. Pressure at the end of the expansion, 21.33 pounds per square inch.

V. *Analysis of the Inertia of the Recorder. Selection of the Spring to be Employed.*—Given the rapidity with which the explosions succeed one another in automobile engines, it is readily understood that the inertia of the moving parts of the recorder will be graphically reproduced (Fig. 144). The effect of this inertia is a function of the weight of the moving parts and of the extent of their travel.

The moving masses are represented by the piston and its rod, the spring and the levers of the parallelogram stylus. The effects due to inertia have been considerably lessened by reducing the weight of the various parts to a minimum. A hollowed piston, a hollowed rod and short and light levers have been adopted. The traditional pencil has been displaced by a silver point which traces its mark upon a metallically coated paper. For the heavy springs with their long travel, light but powerful springs with small amplitudes have been substituted. Since the perfect lubrication of the recorder-cylinder is of great importance, a simple oiling device certain in its action has been adopted. The recess of the piston forms a cup that can be filled with oil whenever the spring is changed.

At each explosion the violent return of the piston splashes oil against the cylinder walls and thus insures perfect lubrication. It should be observed that if the directions given are not followed, particularly in the choice of a spring suitable for each experiment, inertia effects will be produced. These can easily be detected on the record and cannot be confused with the curves which interpret the phenomena occurring in the cylinder of the engine. At a height equal to the end of the piston's stroke, the cylinder of the recorder is provided with a water-jacket which keeps the temperature down to a proper point and prevents the binding of the piston.

The explosion-chamber of automobile engines being rather small in volume, should not be sensibly increased in order that the record obtained may conform as nearly as possible with actual working conditions on the road. In order to attain this end the cylinder of the recorder is so disposed that the piston travels to the height of the connecting-cock. As a result of this arrangement the field of action of the gases is reduced to a minimum. Since these gases have no winding path to follow, they are subjected neither to loss of quantity nor to cold.

CHAPTER XV

THE SELECTION OF AN ENGINE

THE conditions which must be fulfilled both by engines and gas-producers in order that they may industrially operate with regularity and economy have been dwelt upon at some length. Unfortunately it often happens that engines are not installed as they should be, with the result that they run badly and that the reputation of gas-engines suffers unjustly. The use of suction gas-producers in particular caused considerable trouble at first owing to inexperience, so that even now many hesitate to adopt them despite their great economical advantages. The reason assigned for this hesitation is the supposed danger attending their operation.

The factory proprietor who intends to install a gas-engine in his plant is not usually able to appreciate the intrinsic value of one engine when compared with another, or to determine whether the plans for an installation conform with the best practice. The innumerable types of engines offered to him by manufacturers and their agents, each of whom claims to have a better engine than his rivals, plunges the purchaser into hesitation and doubt. Not knowing which engine to select, he usually buys the cheapest. Very often he learns, as time goes by, that his installation is far from being perfect.

Finally he begins to believe that he ought to consult an expert. The author's personal experience has convinced him that eight times out of ten the factory owner who has picked out an engine for himself has not obtained an installation which meets the requirements which the manufacturers of gas-engines should fulfil. Many of these requirements could be complied with were it not for the fact that the manufacturer has dropped certain details which appeared superfluous, but which were in reality very important in obtaining perfect operation. The author therefore suggests that the services of a competent expert be retained by those who intend to install a gas-engine in their plants.

The Duty of a Consulting Engineer.—An expert fills the same office as an architect, and impartially selects the engine best suited to his client's peculiar needs. His examination of the engines offered to him will proceed somewhat according to the following programme:

1. He will first study the installation from the mechanical point of view, and also the local conditions under which that installation is to operate, in order that he may not order an engine too large or too small, or a type incompatible with the foundations at his disposal, or unable to fulfil all the requirements of his client.

2. He will examine the precautions which have been taken to avoid or reduce to a minimum certain inconveniences which attend the operation of explosion-engines.

3. He will draw up specifications, with the terms of which gas-engine makers must comply, so that he can compare on the basis of these specifications the merits of the engines submitted to him.

4. He will prepare an estimate of cost and also a contract which is not couched in terms altogether in the gas-engine maker's favor, and which gives the purchaser important warranties.

5. He will supervise the technical installation of the engine or plant.

6. He will make tests after the engine is installed and see to it that the maker has fulfilled his warranties.

Specifications.—Since engines and gas-producers are constructed for commercial ends, it naturally follows that their manufacturers seek to make the utmost possible profit in selling their installations. Prices charged will necessarily vary with the quality of material employed, the care taken in constructing the engine and generator, the number of apparatus of the same type which are manufactured, the arrangement of the parts and that of the installations. Since there is considerable rivalry among gas-engine builders, selling prices are often cut down so far that little or no profit is left. It is very difficult—indeed impossible—to convince a purchaser that it is to his interest to pay a fair price in order to obtain a good installation, especially when other manufacturers are offering the same installation at a less price with the same warranties. As a result of this state of affairs, engine builders, in order that they may not lose an order, are willing to reduce their prices, hoping to make up in the quality of the workmanship and the material what they would otherwise lose. Often they will deliver an engine too small in size but operating at a higher speed than that ordered; or they will select an old type, or carry out certain details with no great care.

This, to be sure, is not always the case; for there are a few builders of engines who place their reputation above everything else and who would rather lose an order than execute it badly. Others, unfortunately, prefer to have the order at all costs.

By retaining a consulting engineer, all these difficulties are overcome. In the first place, the engineer draws up a scale of prices and specifications which must be complied with in their entirety as well as in all details. Rival engine builders are thus compelled to make their estimates according to the same standard, so that one engine can readily be compared with another with the utmost fairness. In these specifications, penalties will be provided for by the engineer which will be levied if the warranties of the maker are not fulfilled. Otherwise the warranties are worth nothing.

The first consequence of engaging a consulting engineer is to render the matter of cost a secondary one. A factory owner who employs a consulting engineer and pays him for his services, is impelled chiefly by the desire to obtain a good installation which will perform what he expects of it. For that reason necessary sacrifices will be made to comply with the client's wishes.

If the purchaser considers the question of cost most important to him, he need not engage an expert to supervise the installation of his engines. He has simply to pick out the cheapest engine. Unfortunately, however, the money which he will save by such a procedure will be more than compensated for by the trouble which he will later experience when his motor stops or when it breaks down, because it has been cheaply built in the first place.

The advice of a consulting engineer is therefore of importance to the purchaser, because an engine will be installed which will in every way meet his requirements. The gas-engine builder will also prefer to deal with an engineer, because the engineer can appreciate at their true worth good material and good workmanship and place a fair valuation upon them. The specifications of a gas-engine and gas-producer expert are accepted by most engine builders, because an expert will not introduce conditions which cannot be fulfilled. Some manufacturers refuse to consider the conditions imposed by specifications seriously, or else they fix different prices and make tenders on the basis of these with or without specifications. In either case the purchaser may be sure that he is not receiving what he has a right to exact.

Testing the Plant.—When the engine has been selected the consulting engineer supervises its installation, and, after this is completed, carries out tests in order to determine whether or not the guaranteed power and consumption are attained. The methods employed in testing a gas-engine are both complex and delicate. The quality of the gas, the proportions of the elements forming the mixture, the time and the method of ignition, the temperature of the cylinder-walls, the temperature and the pressure of the gas drawn into the cylinder, all these are factors which have a decided bearing upon the results of a test. If these factors be not carefully considered the conclusions to be drawn from the test may be absolutely wrong.

Indicators of any type should not be indiscriminately employed; only those specially designed for gas-engine purposes should be used. Indicator cards are in themselves inadequate, and should be supplemented by the records of explosion-recorders.

The calorific value of the gas should be measured either by the Witz apparatus or by means of any other calorimeter.

In interpreting the diagrams and records some difficulty will be encountered. Sometimes it happens that a particular form of curve is attributed to a cause entirely different from the real one. It happens not infrequently that engineers, whose experience is confined to engines of one make and who have not had the opportunity to make sufficient comparisons, draw such erroneous conclusions from cards.

To recapitulate what has already been said, the testing of gas-engines requires considerable experience and cannot be lightly undertaken. Special instruments of precision are necessary. The author has very often been called upon to contradict the results obtained by experts whose tests have consisted simply in ascertaining the engine power either by means of a Prony brake, or by means of a brake-strap on the fly-wheel. The brake gives but crude results at best; it is a means of control, and not an instrument of scientific investigation.

Something more than the mere power produced by an engine should be ascertained. The tests made should throw some light upon the reasons why that power cannot be exceeded, and show that the necessary changes can be made to cause the engine to operate more economically and to yield energy of an amount which its owner has a right to expect. The indicator and the recorder are testing instruments which clearly indicate discrepancies in operation and the means by which they may be corrected. The tests made should determine whether the power developed is not obtained largely by means of controlling devices which cause premature wearing away of the engine parts.

It is not the intention of the author to describe indicators of the well-known Watt type. It is simply his purpose to call attention to the explosion-recorder which he has devised to supplement the data obtained by means of the indicator.

Explosion-Recorder for Industrial Engines.—The explosion-recorder illustrated in Fig. 145 can be adapted to any ordinary indicator. It is composed of a supporting bracket *B* upon which a drum *T* is mounted. This drum is rotated by a clock-train, the speed of which is controlled by means of a special compensating governor. The entire system is pivotally mounted upon the supporting screw *O*, so that the drum *T*, about which a band of paper is wound, may be

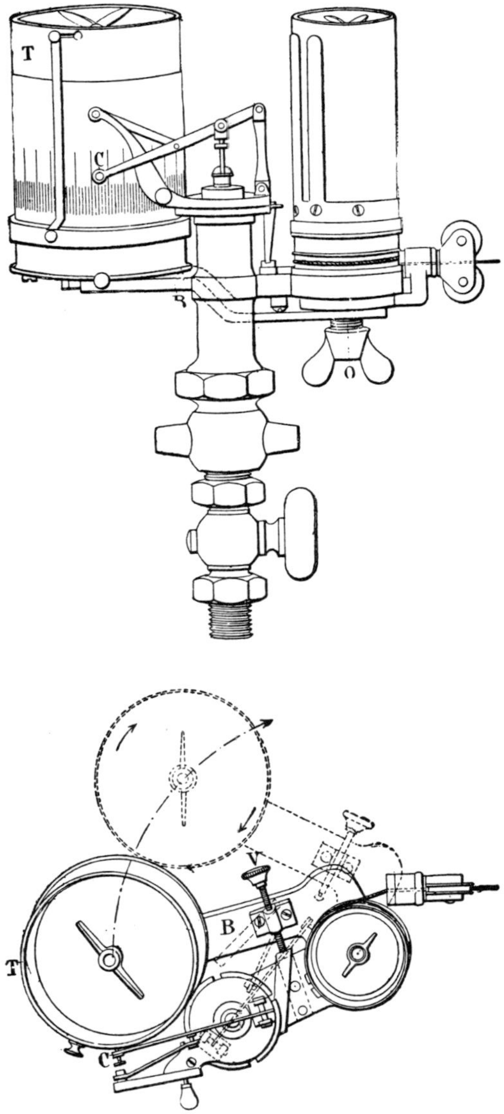

FIG. 145.—Mathot explosion-recorder.

swung against a stylus *C*, which records upon the paper the number and power of the explosions. These explosions are measured according to scale by a spring connected with an indicator. The records obtained disclose for any given cycle the amount of compression as well as the force of the explosion, and render it possible to study the phenomena of expansion, exhaust, and suction. They are, however, inadequate in showing exactly how an engine runs in general. Indeed, in most gas-engines, as well as oil and volatile hydrocarbon engines, each explosion differs from that which follows in character and in power; and it is absolutely essential to provide some means of avoiding these variations. The explosion-recorder gives a graphic record from which the number of explosions can be read, and also the initial pressure of each explosion, the number of corresponding revolutions, the order in which the explosions succeed one another, and consequently the regularity of certain phenomena caused by secondary influences, such as the section of the distributing members, the sensitiveness of the governor, and the like.

The explosion-records can be taken simultaneously with ordinary diagrams. In order to attain this end, the recorder is allowed to swing around the pivot *O*, so that the drum carrying the paper band is brought into engagement, or swung out of engagement with the stylus, as it is influenced by each explosion, thereby leaving its record on the paper. The ordinary diagram may be traced on the drum of the indicator, as it continues to operate in its usual way. Thus the explosion-recorder renders it possible to control the operation of engines, to obtain some idea of the cause of defects and to attribute them to the proper force. Improvements can then be made which will ensure a greater efficiency. A number of records herewith reproduced illustrate the defects in the controlling apparatus and in the construction of certain engines, and also the result of improvements which have been made on the basis of the records obtained. The smaller lines indicate the compression, which is usually constant in engines in which the "hit-and-miss" system of governing is employed, while the larger lines indicate the explosions. These records are only part of the complete data normally drawn on the

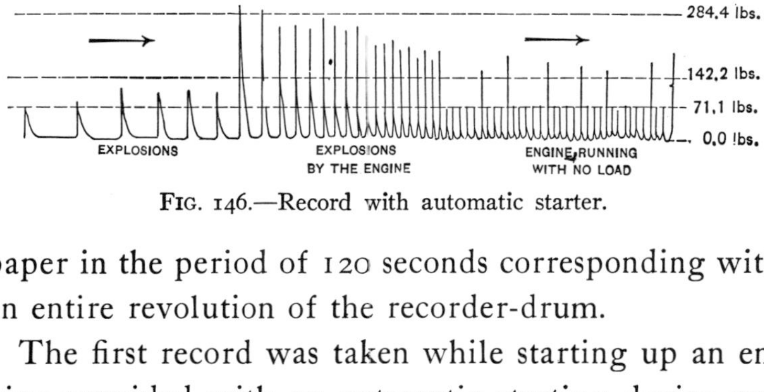

FIG. 146.—Record with automatic starter.

paper in the period of 120 seconds corresponding with an entire revolution of the recorder-drum.

The first record was taken while starting up an engine provided with an automatic starting device and supplied with explosive mixture without previous compression (Fig. 146). The gradual lessening of the distances of the ordinates or lines representing the explosions shows that the speed of the motor was slowly increasing, and also indicates the time which elapsed before the engine was running smoothly. The records that follow (Figs. 147, 148 and 149) show the results which can be obtained with the recorder by correcting the errors due to faults in installing the engine and its

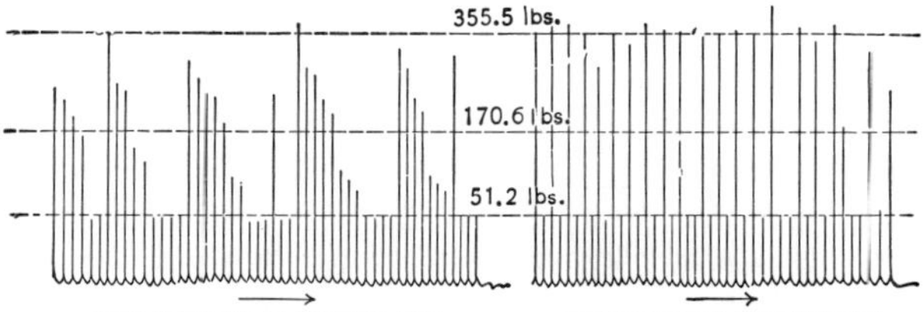

FIG. 147.—Gas-engine running at one-half load.

accessories. The fifth record is particularly interesting because it shows the influence of the ignition-tube on

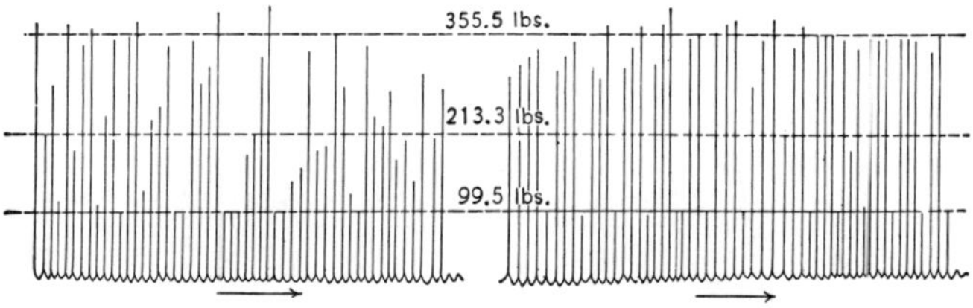

FIG. 148.—Record made after correcting faults.

the power of the deflagration of the explosive mixture (Fig. 150). This record was obtained with an engine

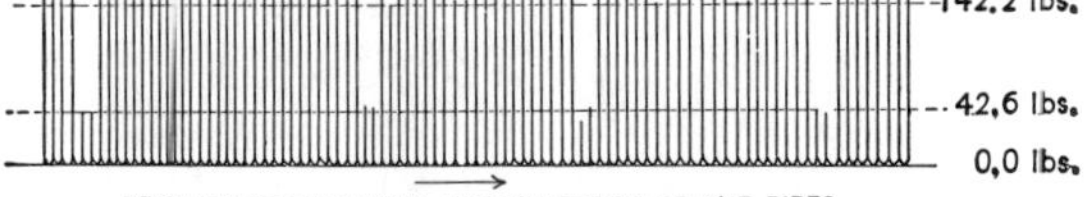

FIG. 149.—Record made after correcting faults.

provided with two contiguous tubes. The communication of each of these tubes with the explosion-chamber could be cut off at will at any moment. The last record (Fig. 151) was obtained at a time when the effective load of the engine was changed at two different intervals. This record shows how regularly the engine was running and how constant were the initial pressures. These pressures, however, which is the case in most engines, manifestly diminish when the explosions succeed one another without idle strokes of the piston. This shows, also, the influence of "scavenging" the products of combustion and the effect it has on the efficiency of explosion-engines.

Analysis of the Gases.—It has already been stated that one of the tests which should be made consists in measuring the calorific value of the gas. Just what the calorific value of the gas may be it is necessary to know in order to obtain some idea of the thermal efficiency of the installation. If a suction gas-producer be employed (an apparatus in which the nature of the gas generated changes at each instant), calorimetrical analyses are indispensable in appreciating the conditions under which a generator operates.

These analyses are made by means of calorimeters which give the calorific value either at a constant pressure or at a constant volume.

Constant-volume instruments give a somewhat weaker record than constant-pressure instruments; but according to Professor Aimé Witz, the inventor of an excellent calorimeter, the constant-volume type is almost indispensable in gaging the efficiency of explosion-engines.

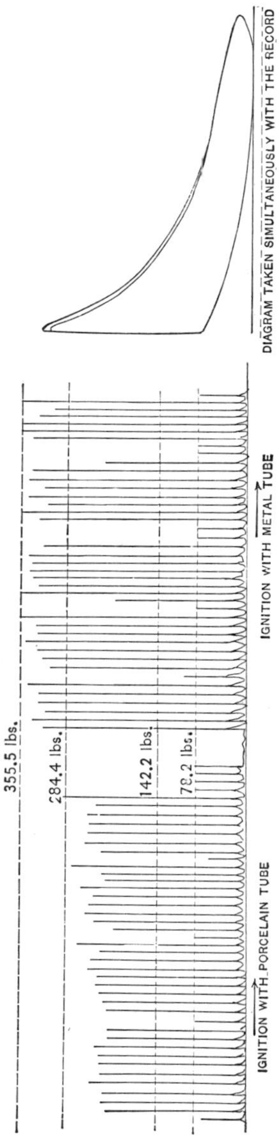

FIG. 150.

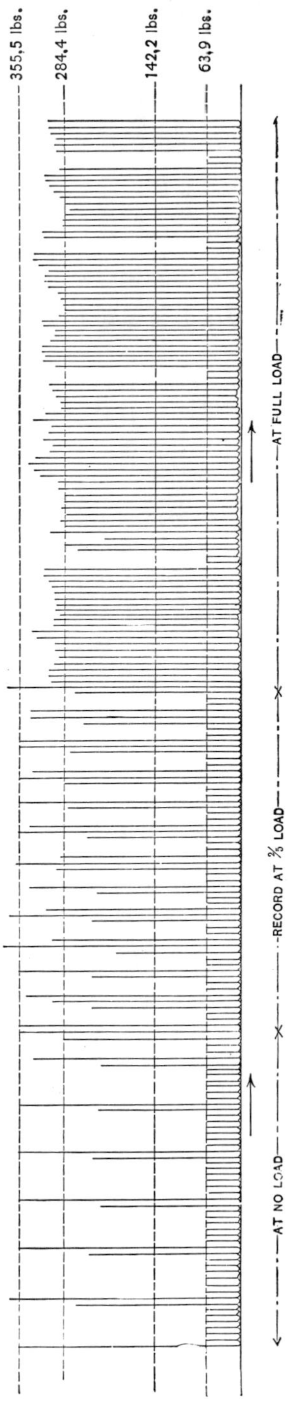

FIG. 151.—Record made when effective load was changed at two different intervals.

The Witz Calorimeter.—The accompanying diagram (Fig. 152) illustrates Professor Witz's instrument. Its elements are a steel cylinder having an interior diameter of 2.36 inches, about a thickness of 0.078 inch and a height of about 3.54 inches, so that its capacity is about 15.1 cubic inches, and two covers screwed on the cylinder to seal it hermetically, oiled paper being used as a washer. The upper cover carries a spark-

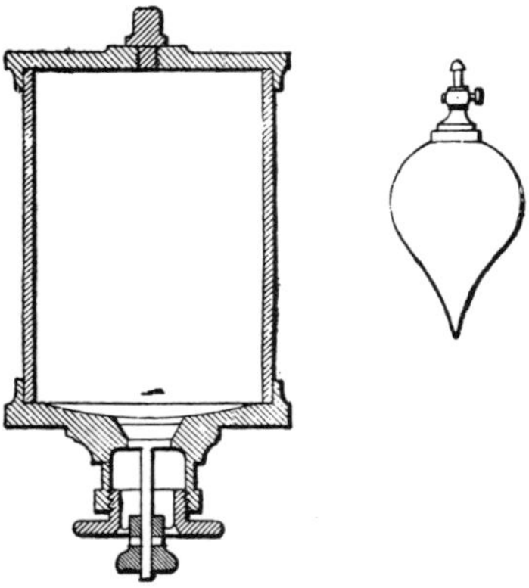

FIG. 152.—The Witz calorimeter.

exciter; the lower cover is provided with a valve which discharges into a cylindrical member 1.06 inches in diameter. This second cover is downwardly inclined at its circumference toward the center to insure complete drainage of the mercury used for charging the calorimeter. All surfaces are nickel plated. The proportions of nickel and of steel are fixed by the manufacturer so as to render it possible to calculate the displacement of the apparatus in water. The calorimeter having been completely filled with mercury is inverted in this liquid in the manner of a test tube. The explosive mixture is then introduced, being fed from a bell in which it has previously been prepared. A rubber tube connects the bell with the instrument. The gas is forced from the bell to the calorimeter by the pressure in the bell. The conical form of the bottom causes the calorimeter to be emptied rapidly and to be refilled completely with explosive gas at a pressure slightly above that of the atmosphere. Equilibrium is re-established by manipulating the valve, during a very short interval, so as to permit the excess gas to escape. During this operation the calorimeter must be maintained in the vertical position shown in the diagram. The atmospheric pressure is read off to one-tenth of a millimeter (0.003936 inches) on a barometer. The temperature of the gas may be taken to be that of the mercury-vessel.

The explosive mixture is prepared in the water reservoir, the glass bulb shown in the accompanying illustration being employed. This bulb is closed at its upper end by means of a cock and is tapered at its lower end. The gas or air enters at the top by means of a rubber tube and gradually displaces the water through the lower end. The bulbs have a volume varying from 200 to 500 cubic centimeters (12 to 30 cubic inches), and the error resulting from each filling of a bulb is certainly less than 15 cubic millimeters (0.0009 cubic inches). The contents are emptied into a bell by lowering the bulb into the water and opening the cock. If seven bulbfuls of air be mixed with one bulbful of gas, an explosive mixture of 1 to 7 is produced, this being the proportion commonly employed for street-gas. For producer-gases the preferred proportion is 1 to 1, oxygen being often added to the air in order to insure complete combustion.

The calorimeter, after having been filled, is placed in a vessel containing a liter (1.7598 pints) of water so that it is completely immersed. A spark is then allowed to pass. The explosion is not accompanied by any noise; the temperature rises a fixed number of degrees, so that the quantity of heat liberated can easily be computed. Each division of the thermometer is equal to 0.01502 C. The scale reading is minute, each interval being divided by ten, so that readings to the 1,500th part of a degree can be taken.

It should be observed that the mixture generated in the reservoir is saturated with water vapor at the temperature of the reservoir. Consequently, the vapor generated by the explosion must condense in the calorimeter if the final temperature of the calorimeter is the same as that of the water reservoir. If, on the other hand, the temperature be slightly different, a correction must be made; but the error is negligible for differences in temperature of from 2 to 3 degrees C. (3.6 to 5.4 degrees F.). This, however, is never likely to occur if the operation is conducted under favorable conditions.

This apparatus is exceedingly simple and practical. It does not require the manipulation of a pump. The pressure of the mixture is read off on the barometer; the calorimeter is entirely immersed in the water of the outer vessel, so that all corrections of doubtful accuracy are obviated. The method requires but a very slight correction for temperature. Air, alone or mingled with oxygen, or a mixture of air and oxygen, can be easily tested with.

Maintenance of Plants.—If it should be necessary to retain a consulting engineer to install an engine capable of filling all requirements, it is also necessary to select a careful attendant in order that the engine may be kept in good condition. It is a rather widespread belief that a gas-engine can be operated without any care or inspection. This belief is all the more prevalent because of the employment of street-gas engines, which, by reason of their simplicity of construction and regularity of fuel supply, often run for several hours, and even for an entire day, without any attention whatever. But this negligence, particularly in the case of engines driven from producers, is likely to produce disastrous results. Although engines of this type do not require constant inspection during operation, still they require some attention in order that the speed may be kept at a fixed number of revolutions. Moreover, the care of the engine, the cleaning of the valves and of the various parts which are likely to become dirty, and the examination and cleaning of pipes, should be accomplished with great care and at regular intervals. This task should be entrusted only to a man of intelligence. A common workman who knows nothing of the care with which the parts of an engine should be handled is likely to do more harm than good.

The factory owner who follows the instructions which have been given in this book will avoid most of the stoppages and the trouble incurred in engine and generator installations, and may count upon a steadiness of operation comparable with that of a steam-engine.

TEST OF A "STOCKPORT" GAS-ENGINE WITH DOWSON PRESSURE GAS PLANT

Made by R. Mathot at the Works of the "Union Electrique" Cie, Brussels, June 27, 1901

Piston Diameter: 15½". Piston stroke, 22". Normal number of revolutions, 210.

1. Calorific value of the coal	12750 B.T.U.
2. Nature and origin of fuel: Anthracite coal of Charleroi (Belgium).	
3. Cost of fuel per ton at the mine	$5.50
4. Cost of fuel per ton at the plant	$6.39
5. Fuel consumption per hour in the generator .	46.3 lbs.
6. Fuel consumpton per hour in the boiler .	7 lbs.
7. Proportion of ash in the coal	6 per cent.
8. Weight of steam at 66 lbs. generated per hour	42.7 lbs.
9. Average brake horse-power	53 B.H.P.
10. Fuel consumption for gas per B.H.P. per hour	0.875 lbs.
11. Fuel consumption for steam per B.H.P. per hour	0.133 lbs.
12. Total fuel consumption	1.008 lbs.
13. Steam consumption at 66 lbs. pressure . .	0.81 lbs.
14. Gas pressure at the engine	1⅜ inches
15. Weight of water per B.H.P. per hour for cooling the cylinder entering at 68° F. and leaving at 105° F.	51.5 lbs.

16.	Corresponding heat absorbed in cooling .	1970 B.T.U.
17.	Average initial explosive pressure on piston	324 lbs.
18.	Average pressure on piston per square inch	72 lbs.
19.	Average indicated horse-power with 85 per cent. misses	92.5 I.H.P.
20.	Corresponding mechan'cal efficiency . .	84 per cent.
21.	Corresponding electric load	31.950 K.W.
22.	Cost of B.H.P. per hour in anthracite . .	\$0.0029
23.	Cost of kilowatt per hour in anthracite . .	\$0.0048
24.	Electric power generated per B.H.P. . .	602.8 W.
25.	Thermal efficiency at 53 B.H.P. with 85 per cent. explosions	18.5 per cent.

TEST OF A 20 H.P. WINTERTHUR ENGINE

With Winterthur Suction-Producer made by R. Mathot at Winterthur, June 4 and 5, 1902

DATA OF TESTS WITH ILLUMINATING GAS AND WITH FUEL GAS

Dimensions of Winterthur Engine—Piston diameter: 10⅜″. Stroke: 16⅞″. Compression: 177 pounds per square inch. Regulation: hit and miss. Ignition: electro-magnetic. Fly-wheel: normal, with external bearing. Lubrication of piston: with oil-pump. Of main bearings, with rings (as in dynamos).

FULL LOAD WITH STREET-GAS

1.	Number of revolutions per minute . . .	200
2.	Corresponding number of explosions . .	96 per cent.
3.	Net load on brake	120 lbs.
4.	Corresponding effective power	22 B.H.P.
5.	Mean initial explosive pressure on piston per square inch	455 lbs.
6.	Average pressure on piston per square inch	78 lbs.
7.	Gas consumption per B.H.P. at 24° C. and 721 mm. mean pressure	15.5 cubic feet
8.	Gas consumption per B.H.P. reduced to 0° C. and 760 mm. mean pressure . .	13.5 cubic feet

HALF LOAD WITH STREET-GAS

9.	Number of revolutions per minute . . .	204
10.	Corresponding number of explosions . .	60 per cent.
11.	Net load on brake	60 lbs.
12.	Corresponding effective power	11.6 B.H.P.
13.	Gas consumption per B.H.P. per hour at 24° C. and 721 mm. mean pressure . .	21 cubic feet
14.	Gas consumption per B.H.P. per hour at 0° C. and 760 mm. mean pressure . .	18.3 cubic feet

RUNNING WITH NO LOAD WITH STREET-GAS

15.	Number of revolutions per minute . . .	206
16.	Corresponding number of explosions . .	22 per cent.
17.	Total gas consumption per hour at 24° C. and 721 mm. mean pressure	106 cubic feet
18.	Maximum calorific power of gas per cubic foot	598 B.T.U.
19.	Thermal efficiency with 96 per cent. explosions	31 per cent.
20.	Mechanical efficiency with 96 per cent. explosions	82 per cent.
21.	Temperature of water at the jacket-inlet .	75 degs. F.
22.	Temperature of water at the jacket-outlet	130 degs. F.
23.	Compression per square inch on piston surface	178 lbs.
24.	Pressure after expansion	37 lbs.

TEST OF WINTERTHUR PLANT WITH PRODUCER-GAS

1. Nature of fuel. Belgian anthracite, "Bonne Esperance et Batterie"; size, ¾ inch.
2. Chemical composition: Carbon, 86.5 per cent.; hydrogen, 3.5 per cent.; oxygen and nitrogen, 4.65 per cent.; ash, 5.35 per cent.
3. Calorific value per pound of coal . . . . 14200 B.T.U.

4.	Net calorific value per pound of fuel . .	15050 B.T.U.
5.	Price of anthracite delivered at the plant .	$3.50 per ton
6.	Number of revolutions of engine per minute	200
7.	Corresponding number of explosions . .	91 per cent.
8.	Load on brake	106 lbs.
9.	Corresponding effective horse-power . .	20.2 B.H.P.
10.	Fuel consumption at the generator per hour	16.4 lbs.
11.	Fuel consumed per B.H.P. per hour . .	0.81 lbs.
12.	Proportion of ash resulting from the tests .	6 per cent.
13.	Mean initial explosive pressure per square inch	419.5 lbs.
14.	Average pressure on piston per square inch	72.5 lbs.
15.	Indicated horse-power with 91 per cent. explosions	25.4 I.H.P.
16.	Mechanical efficiency	79 per cent.
17.	Thermal efficiency at the producer . . .	22 per cent.
18.	Water consumption per hour in the scrubber	66 gals.
19.	Cost per B.H.P. per hour in anthracite .	62 gals.

TEST OF A 60 B.H.P. GAS-ENGINE, TYPE G 9, WITH A SUCTION-GAS PLANT OF THE GASMOTOREN FABRIK DEUTZ

(Made at Cologne, March 15, 1904, by R. Mathot.)

DATA OF THE TESTS

Diameter of Piston = 16.5″. Piston Stroke = 18.9″

FULL LOAD

1.	Average number of revolutions per minute	188.66
2.	Corre ponding effective work	65.11 B.H.P.
3.	Average compression per square inch .	176 lbs.
4.	Average initial explosive pressure per square inch	397 lbs.
5.	Average final expansion pressure . . .	25 lbs.
6.	Vacuum at suction	4.4 lbs.
7.	Average pressure on piston	81 lbs.
8.	Corresponding indicated horse-power . .	77 I.H.P.

FUEL

9. Nature of fuel: Anthracite coal 0.4″ to 0.8″
10. Origin: Coalpit of Ze he, Morsbach at Aix-la-Chape le.
11. Chemical composition of coal:

Carbon	83.22 %	Sulphur . . .	0.44 %
Hydrogen . . .	3.31 %	Ash	7.33 %
Nitrogen and Oxygen	3.01 %	Water . . .	2.69 %

12. Calorific value . . . . 13650 B.T.U.

GAS

13. Chemical composition of gas:

Carbonic acid .	6.60 %	Methane . . .	0.57 %
Oxygen . . .	0.30 %	Carbon monoxide	24.30 %
Hydrogen . .	18.90 %	Nitrogen . . .	49.33 %

14. Calorific value of gas, combination water, at 59° F. constant vo ume reduced to 32° F. and a:mospheric pressure . . . 140 B.T.U.

TEMPERATURES

Engine

15. Cooling water at the inlet of the cylinder-head . . . . 55.4 deg. F.
Temperature at the outlet . . . . 109.5 deg. F.
16. Temperature at outlet of cylinder . . . 127.5 deg. F.

Gas-Generator

17. Temperature of water in the vaporizer . . 158.3 deg. F.

EFFICIENCIES AND CONSUMPTION

18. Mechanical efficiency . . . . 84.6 %
19. Gross consumption of coal per B. H. P. per hour . . . . 0.86 lbs.
20. Thermal efficiency in proportion to the effective work and the gross consumption of coal in the gas-generator . . . . 24.3 %

HALF LOAD

WORK

1.	Average number of revolutions per minute	195.5
2.	Corresponding effective work	33.85 B.H.P.
3.	Corresponding average compression . .	125 lbs.
4.	Average initial explosive pressure . . .	258 lbs.
5.	Average final expansion	18 lbs.
6.	Vacuum at suction	6.8 lbs.
7.	Average mean pressur on piston . . .	46.2 lbs.
8.	Corresponding indicated power	45. I.H.P.
9.	Speed variation between full and half load	3.5 %

CONSUMPTION

10.	Gross consumption of coal per B.H.P. per hour	1.155 lbs.

RUNNING WITH NO LOAD

1.	Average number of revolutions per minute	199
2.	Minimum corresponding compression . .	95.55 lbs.
3.	Average initial explosive pressure . . .	220 lbs.
4.	Average final expansion	0 lbs.
5.	Vacuum at suction	8.8 lbs.
6.	Average pressure on piston	11.2 lbs.
7.	Corresponding indicated horse-power . .	11 I.H.P.
8.	Speed variation between full load and no load	5.2 %

TEST OF A GAS PLANT OF A FOUR-CYCLE DOUBLE-ACTING ENGINE OF 200 H.P. AND A SUCTION-PRODUCER IN THE WORKS OF THE GASMOTOREN FABRIK DEUTZ, COLOGNE

March 14 and 15, 1904, by Me rs. A. Witz, R. Mathot, and de Herbais

DATA OF THE TESTS

Piston Diameter: 21¼″. Stroke: 27 9/16″. Diameter of Piston-Rods: front, 4¾″; rear, 4 5/16″

ENGINE

Full Load Tests

1. Average number of revolutions per minute 151.29 and 150.20
2. Corresponding effective load . 214.22 B.H.P. and 222.83 B.H.P.
3. Duration of the tests . . . . . . 3 hours and 10 hours
4. Average temperature of water after cooling the piston . . . . 117.5 deg. F.
5. Average temperature of water after cooling the cylinder and valve-seats . . . 135 deg. F.
6. Water consumption per hour for cooling the piston . . . . 39 gal.

PRODUCER

7. Nature and Origin of Fuel: Anthracite coal "Bonne-Esperance et Batterie" Herstal, Belgium.
8. Calorific value of fuel . . . . 14650 B.T.U.
9. Consumption of fuel per hour (plus 53 lbs. on the night of the 14th for keeping the generator fired during 14 hours, the engine being stopped) . . . . 199 lbs.—160 lbs.
10. Water consumption per hour in the vaporiser . . . . 14.2 gals.
11. Water consumption per hour in the scrubbers . . . . 318 gals.
12. Average temperature of gas at the outlet of the generator . . . . 558 deg. F.
13. Average temperature of gas at the outlet of the scrubbers . . . . 62.5 deg. F.

EFFICIENCIES

14. Gross consumption of coal per B.H.P. per hour . . . . 0.927 lbs.—0.720 lbs.
15. Consumption of coal per B.H.P. after deduction of the water . . . . . . 0.907 lbs.—0.705 lbs.

16. Thermal efficiency relating to the effective H.P. and to the dry coal consumed in the generator	19 %—24.4 %
17. Water consumption per B.H.P. hour:	
For the cylinder, stuffing-boxes and valve-seat jackets	4.65 gals.
For the piston and piston-rods . . .	1.75 gals.
For the vaporizer	0.0655 gals.
For washing the gas in the scrubbers .	1.42 gals.
18. Water converted in steam per lb. consumed in the generator	0.193 gals.

INDEX

C

D

E

F

H

I

J

L

V

W